Collins New Primary Maths

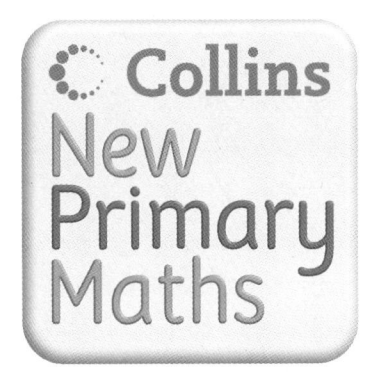

Homework Pack 6

Series Editor: Peter Clarke

Authors: Jeanette Mumford, Sandra Roberts, Andrew Edmondson

William Collins' dream of knowledge for all began with the publication of his first book in 1819. A self-educated mill worker, he not only enriched millions of lives, but also founded a flourishing publishing house. Today, staying true to this spirit, Collins books are packed with inspiration, innovation and practical expertise. They place you at the centre of a world of possibility and give you exactly what you need to explore it.

Collins. Freedom to teach.

Published by Collins
An imprint of HarperCollinsPublishers
77 – 85 Fulham Palace Road
Hammersmith
London
W6 8JB

Browse the complete Collins catalogue at
www.collinseducation.com

© HarperCollinsPublishers Limited 2007

10 9 8 7 6 5 4

ISBN-13 978 0 00 722048 9

The authors assert their moral rights to be identified as the authors of this work

Any educational institution that has purchased one copy of this publication may make unlimited duplicate copies for use exclusively within that institution. Permission does not extend to reproduction, storage within a retrieval system, or transmittal in any form or by any means, electronic, mechanical, photocopying, recording or otherwise, of duplicate copies for loaning, renting or selling to any other institution without the permission of the Publisher.

British Library Cataloguing in Publication Data
A Catalogue record for this publication is available from the British Library

Cover design by Laing&Carroll
Cover artwork by Jonatronix Ltd
Internal design by Steve Evans and Mark Walker Design
Illustrations by Steve Evans and Mark Walker
Edited by Jean Rustean
Proofread by Ros Davies

Printed and bound by Martins the Printers, Berwick-upon-Tweed

FSC Mixed Sources
Product group from well-managed forests and other controlled sources
www.fsc.org Cert no. SW-COC-1806
© 1996 Forest Stewardship Council

Contents

Unit A1

HCM 1   Differences
Find the difference between a positive and a negative number

HCM 2   Making decimals
Order decimals to two places

HCM 3   Revising multiplication and division
Use known number facts for mental multiplication and division involving decimals

HCM 4   More brackets
Use multiplication and division facts and brackets to solve number calculations

Unit B1

HCM 5   Find the calculation
Use place value and multiplication and division facts to work out other facts involving decimals

HCM 6   Square numbers
Recognise squares of numbers to 12 x 12

HCM 7   Puzzling triangular numbers
Describe and explain sequences, patterns and relationships

HCM 8   Counting patterns
Recognise and extend number sequences

HCM 9   Constructing kites
Use mathematical vocabulary to describe the features of a 2-D shape

HCM 10   Investigating midpoints
Make and draw shapes accurately; Describe and explain relationships

Unit C1

HCM 11   All square measuring
Read and interpret scales accurately

HCM 12   DIY ranges and modes
Find the mode and range

HCM 13   Documentary bar chart
Represent data in different ways and understand its meaning

HCM 14   Time for homework
Find the mode and range; Represent data in different ways and understand its meaning

Unit D1

HCM 15   Rocky Mountain motoring
Measure and calculate using imperial units (miles and kilometres)

HCM 16   Finding perimeters
Calculate the perimeter of simple compound shapes that can be split into rectangles

HCM 17   Area workout
Calculate the area of a shape formed from rectangles

HCM 18   Making calculations
Use efficient written methods to add whole numbers and decimal numbers

Contents

Unit E1

HCM 19	Colouring fractions	
	Understand improper fractions and mixed numbers	
HCM 20	Sharing pizzas	
	Relate fractions to division	
HCM 21	Percentages at home	
	Understand percentages as the number of parts in every 100	
HCM 22	Decimal subtraction	
	Use efficient written methods to subtract decimals	
HCM 23	Multiplication methods	
	Approximate first. Use efficient written methods to multiply whole numbers	
HCM 24	Design a snowman	
	Solve a problem and check that the answer is correct	

Unit A2

HCM 25	Decimals decimals	
	Order decimal numbers with up to three places	
HCM 26	Money problems	
	Solve multi-step problems	
HCM 27	Multiplying mentally	
	Multiply mentally with whole numbers and decimals	
HCM 28	Family holiday	
	Choose and use appropriate strategies to solve problems	

Unit B2

HCM 29	Multiplication and division facts	
	Use table facts to work out related facts	
HCM 30	Tests of divisibility (1)	
	Know and apply simple tests of divisibility	
HCM 31	Using square numbers	
	Know square numbers to 12 x 12	
HCM 32	Prime numbers	
	Recognise prime numbers less than 100	
HCM 33	Co-ordinates of reflections	
	Recognise where a shape will be after reflection	
HCM 34	Bisecting diagonals	
	Classify quadrilaterals	

Unit C2

HCM 35	Prehistoric weights	
	Record units of measure (mass)	
HCM 36	Mean and median	
	Find the mean and median of a set of data	
HCM 37	Beach time	
	Find the median, mean and range; Represent data in different ways and understand its meaning	
HCM 38	Baked pie charts	
	Answer questions using data from pie charts	

Contents

Unit D2

HCM 39 All the nines supermarket
Use a calculator to solve problems with several steps

HCM 40 Two-step translations
Use co-ordinates and translate shapes on grids

HCM 41 Measuring angles with set squares
Use a protractor to measure and draw acute and obtuse angles to the nearest degree

HCM 42 Inverse checking
Use inverse operations to check results

Unit E2

HCM 43 Cancelling fractions
Simplify fractions by cancelling common factors

HCM 44 Fraction wheels
Find fractions of whole number quantities

HCM 45 Percentages of the day
Find percentages of whole number quantities

HCM 46 Dominoes
Find equivalent percentages, decimals and fractions

HCM 47 Patterns, ratio and proportion
Solve simple problems using ratio and proportion

HCM 48 Mystic rose
Tabulate systematically the information in a problem or puzzle

Unit A3

HCM 49 Order the decimals
Order decimals with up to three decimal places

HCM 50 Making calculations with decimals
Use efficient written methods to add decimals

HCM 51 Recording division
Approximate first. Use efficient written methods to divide whole numbers

HCM 52 Long division
Use efficient written methods to divide whole numbers

Unit B3

HCM 53 Multiplying mentally by 11
Use multiplication facts to work out mentally other facts

HCM 54 Tests of divisibility (2)
Use approximations, inverse operations and tests of divisibility to estimate and check results

HCM 55 Multiples and consecutive numbers
Describe and explain sequences and relationships

HCM 56 Agent P's puzzle
Find which numbers less than 100 are prime

HCM 57 Making a dodecahedron
Make shapes with increasing accuracy

HCM 58 Grid patterns
Make and draw shapes with increasing accuracy

Contents

Unit C3

HCM 59   Converting capacities
Convert smaller units to larger units and vice versa

HCM 60   Currency converter
Use a graph to convert money

HCM 61   Musical pie charts
Represent data in different ways and understand its meaning

HCM 62   Holiday bar charts
Represent data in different ways and understand its meaning

Unit D3

HCM 63   Litres and gallons
Know imperial units (gallons); Know rough equivalents of litres and gallons

HCM 64   Pinboard areas
Calculate the area of a shape formed from rectangles

HCM 65   Finding surface areas
Find the surface area of a cuboid

HCM 66   Checking calculations
Use approximations and inverse operations to estimate and check results

Unit E3

HCM 67   Decimal fraction pairs
Find equivalent decimals and fractions

HCM 68   Percentages
Express one amount as a percentage of another

HCM 69   Addition and subtraction
Use efficient written methods to add and subtract whole numbers and decimals

HCM 70   Ratio investigation
Solve simple problems involving ratio

HCM 71   Solving word problems
Choose and use appropriate number operations to solve word problems involving money

HCM 72   Reflective bunting
Record the information in a problem or puzzle

Differences

- Find the difference between a positive and negative number

1. Write two numbers with a difference of 6 in each box.
At least one of the numbers must be a negative number.

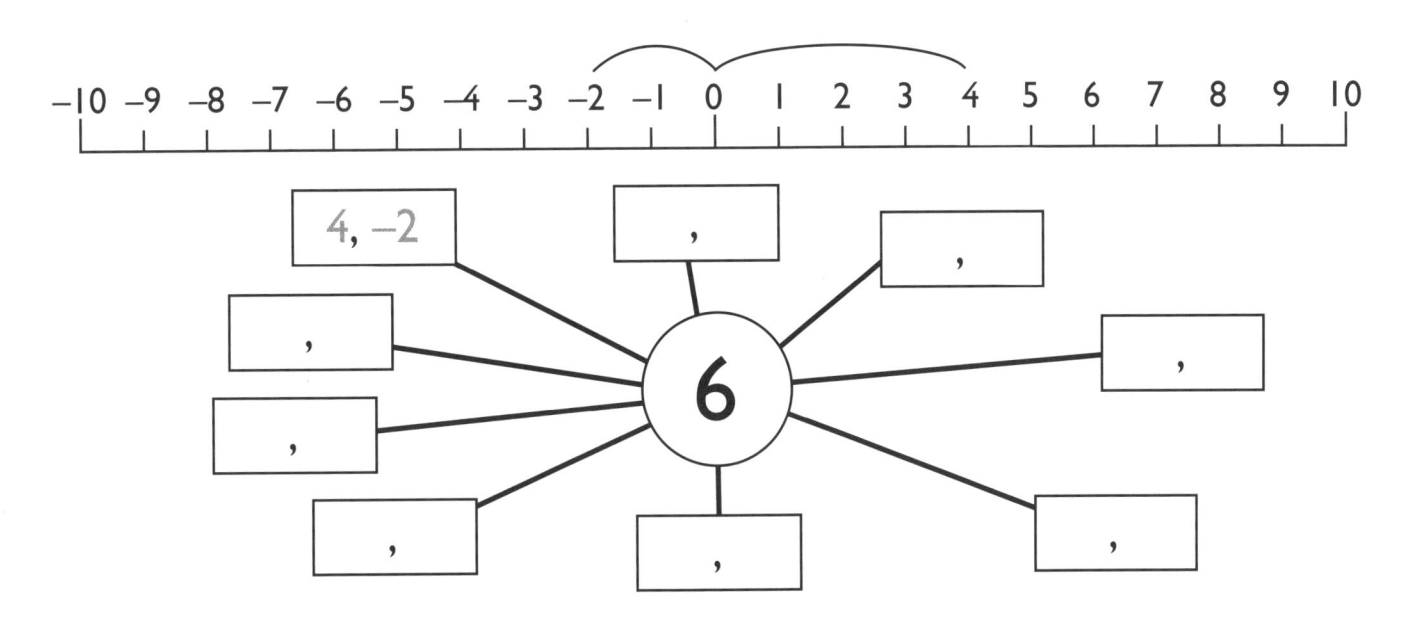

−10 −9 −8 −7 −6 −5 −4 −3 −2 −1 0 1 2 3 4 5 6 7 8 9 10

4, −2

(centre: 6)

2. Write two numbers with a difference of 13 in each box.
At least one of the numbers must be a negative number.

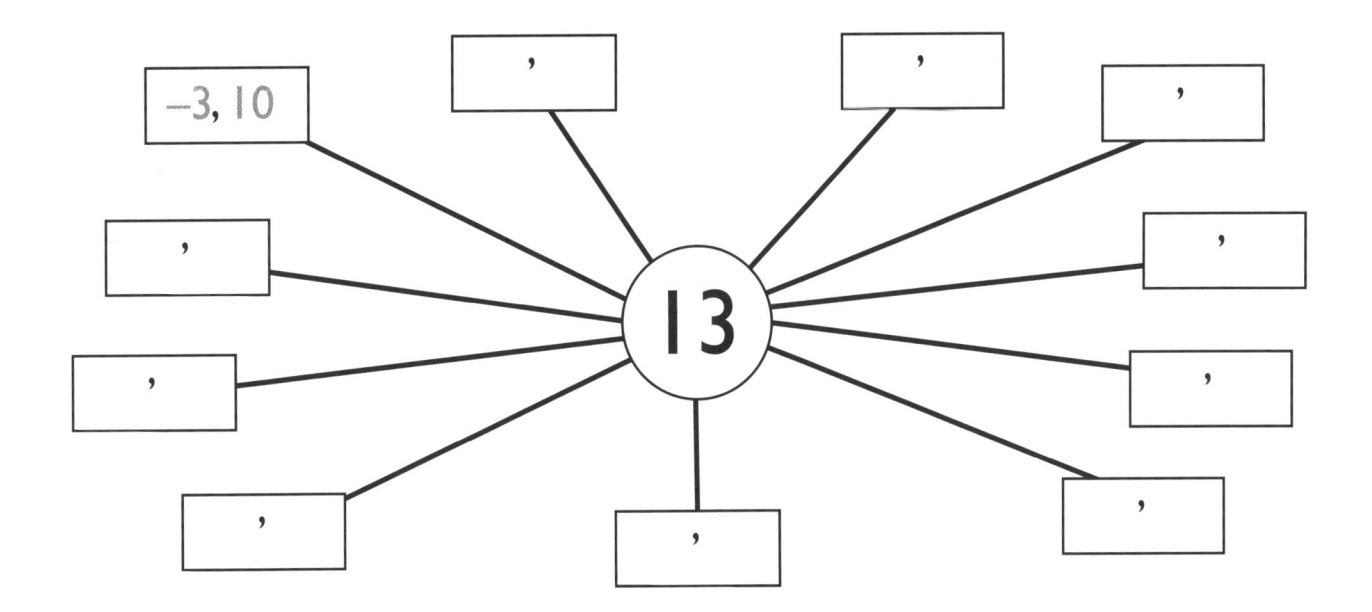

−3, 10

(centre: 13)

▲ Look at the pairs of numbers you wrote with a difference of 13. On the back of this sheet, write a calculation for each of your differences, e.g. −3, 10 → −3 + 13 = 10.

Making decimals

- Order decimals to two places

■ Use the digits on the cards to make decimal numbers to one decimal place. When you have made your numbers, put them in order starting with the smallest.

Decimals to one place

In order

Smallest

Largest

[6] [3] [8] [1]

◤ Use the digits on the cards to make numbers to two decimal places. When you have made your numbers, put them in order, smallest to largest. In the third column, write a number that comes **between** each of your numbers.

[0] [9] [2] [4] [7]

Decimals to two places

In order Smallest	Numbers in between
Largest	

© HarperCollinsPublishers Ltd 2007

Revising multiplication and division

- Use known number facts for mental multiplication and division involving decimals

Buy these items.
Work out the answers to these in your head.

a) 10 @ £1·27 each

b) 10 @ £1·75 each

c) 2 @ £0·95 each

d) 2 @ £0·78 each

e) 5 @ £0·90 each

f) 10 @ £0·86 each

g) 5 @ £6·50 each

h) 100 @ £3·67 each

Buy the amounts of each item shown. Work out the answers in your head.

a) Yo-Yo £1·37

Buy	2	10	5	100	50
× £1·37					

b) Trainers £48

Buy	2	10	5	100	500
× £48					

c) Hi-Fi £79

Buy	10	5	20	100	500
× £79					

d) Books £4·68

Buy	100	200	1000	500	250
× £4·68					

Y6 A1 L7
HCM 3

© HarperCollinsPublishers Ltd 2007

More brackets

- Use multiplication and division facts and brackets to solve calculations

2 3 5

Use the digits 2, 3 and 5 to complete the calculations.

a (△ × △) + △ = 13
b (△ × △) + △ = 17
c △ × (△ + △) = 16
d △ × (△ + △) = 25
e (△ × △) + △ = 11
f (△ − △) × △ = 4
g △ × (△ − △) = 9
h △ × (△ + △) = 21

Remember

Always divide or multiply before you add or subtract unless there are (BRACKETS).

If there are brackets, do what is inside the brackets FIRST!

Find the answers to these.

a 6 × 7 + 5 + 9 =
b 3 × (8 + 4) − 9 =
c 7 × 4 + 5 + 8 =
d 7 × (4 + 5) + 8 =
e (6 × 6) + (8 × 6) =
f 64 ÷ 8 × 7 =
g 100 − (9 × 8) + 16 =
h (56 ÷ 8) × 9 =
i (7 × 7) + (9 × 6) =
j (72 ÷ 8) × 12 =
k 48 ÷ 4 + (7 × 8) =
l 24 + (9 × 9) =

Find the missing number to complete the number sentences.

a 5 × ☐ + 32 = 72
b 5 × (☐ + 32) = 200
c (9 × 7) + (☐ × 3) = 90
d 36 ÷ ☐ × 8 = 72
e 6 × 8 + ☐ = 83
f (48 − ☐) × 5 × 2 = 240
g 3 × 9 + 6 × ☐ = 81
h (7 × ☐) − (14 + 9) = 33
i 49 − (19 + ☐) = 16
j 54 ÷ 9 × ☐ = 36
k 12 + (☐ × 12) = 60
l 16 × ☐ ÷ 4 + 4 = 8
m 6 × 2 × 3 + ☐ = 72
n (4 × 9) + (☐ × ☐) = 42
o 56 ÷ ☐ × 7 = 49

Y6 A1
HCM

Y6 B1 L1
HCM 5

Find the calculation

- Use place value and multiplication and division facts to work out other facts involving decimals

Instructions (for 2 players)

- Shuffle the cards. Place them face down on the table.
- Take turns to choose a card, e.g. 4.
- Use the number to make a multiplication fact. The answer needs to match a number on the board, i.e. 4 × 0·6 = 2·4.
- Cross out the number on the board.
- If a number is already crossed out, miss a go.
- The winner is the first person to cover 3 numbers in a row vertically, horizontally or diagonally.
- The player with the most crosses on the board at the end is the winner.

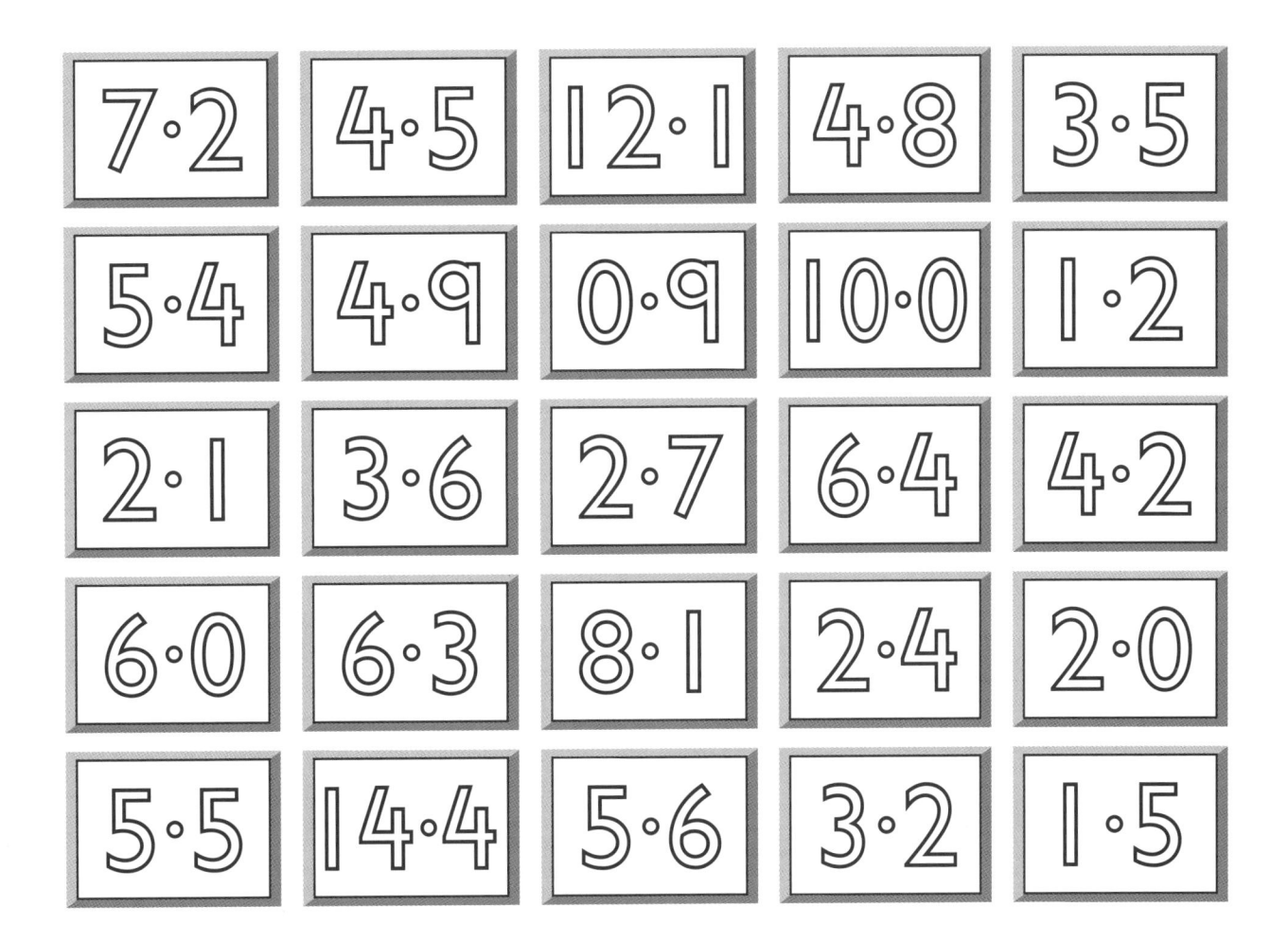

You need:

- 1-12 number cards or Playing Cards where Ace is 1, Jack is 11 and Queen is 12 (and Kings have been removed)
- 2 pens (a different colour per person)

7·2	4·5	12·1	4·8	3·5
5·4	4·9	0·9	10·0	1·2
2·1	3·6	2·7	6·4	4·2
6·0	6·3	8·1	2·4	2·0
5·5	14·4	5·6	3·2	1·5

© HarperCollins*Publishers* Ltd 2007

Name _____ Date _____

Square numbers

- Recognise squares of numbers to 12 x 12

Complete the tables.

a

Counting numbers	1	2	3	4	5	6	7	8	9	10	11	12
Square numbers		4										

b

1^2		7^2	10^2		9^2		3^2		4^2			12^2
	25			4		64		100		121	36	

Use your knowledge of square numbers to work out the answers to these questions.

1. **a** $3^2 + 9 =$ ☐   **b** $6^2 + 8 =$ ☐   **c** $4^2 + 12 =$ ☐

   d $10^2 - 17 =$ ☐   **e** $8^2 + 33 =$ ☐   **f** $7^2 + 10^2 =$ ☐

   g $11^2 - 48 =$ ☐   **h** $5^2 + 17 =$ ☐   **i** $11^2 - 10^2 =$ ☐

2. **a** $3^2 + 5^2 =$ ☐   **b** $4^2 + 6^2 =$ ☐   **c** $9^2 + 8^2 =$ ☐

   d $12^2 + 5^2 =$ ☐   **e** $9^2 - 2^2 =$ ☐   **f** $11^2 + 10^2 =$ ☐

   g $11^2 - 6^2 =$ ☐   **h** $5^2 + 8^2 =$ ☐   **i** $12^2 - 10^2 =$ ☐

Puzzling triangular numbers

- Describe and explain sequences, patterns and relationships

Example
9 = 3 + 6
14 = 1 + 3 + 10

Every whole number is the sum of no more than 3 triangular numbers.

1 a Complete this sequence of triangular numbers.

1, 3, 6, 10, ..., ..., ..., ..., ..., ...

b Write these whole numbers using no more than 3 triangular numbers.

Whole number	Triangular numbers
12	
14	
15	
18	

1 Investigate the statement above for these numbers.

Whole number	Triangular numbers
17	
29	
31	
46	
58	
67	

2 Find 6 numbers between 10 and 50 which will fulfil the gypsy's prediction.

Find numbers which can be made from more than one set of triangular numbers and you will be lucky! I can see
19 = 1 + 3 + 15 or 3 + 6 + 10.

Name _____ Date _____

Y6 B1
HCM

Counting patterns

- Recognise and extend number sequences

■ Complete each number sentence.

1
a – 50 + 25 =
b – 45 + 15 =
c – 99 + 11 =
d – 48 + 12 =
e –275 + 25 =

2
a 152 – 12 =
b 391 – 19 =
c 684 – 21 =
d 395 + 15 =
e 691 + 11 =

3
a 372 + 12 + 12 =
b 146 + 11 + 11 =
c 275 + 25 + 25 =
d 358 + 19 + 19 =
e 586 + 21 + 21 =

● An underwater diver is searching for related fish. He is hoping to find multiples of 15 fish, multiples of 19 fish and multiples of 21 fish.

You need:
- colouring materials

1 Colour each set of multiples in a different colour.

2 Write the numbers in the correct sequence below.

Fish numbers: 114, 63, 150, 45, 90, 38, 147, 57, 126, 135, 84, 133, 189, 171, 76, 120, 75, 95, 210, 60, 168, 152, 105

Multiples of 15 →
Multiples of 19 →
Multiples of 21 →

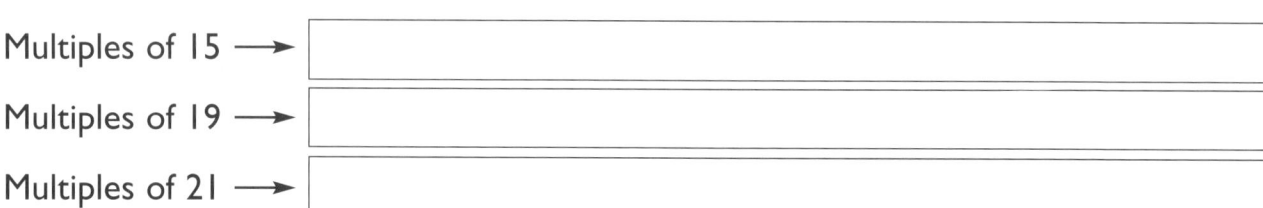

▲ Look at the three sequences above. On the back of this sheet write the next five numbers in each sequence.

© HarperCollinsPublishers Ltd 2007

Y6 B1 L11
HCM 9

Name _____ Date _____

Constructing kites

- Use mathematical vocabulary to describe the features of a 2-D shape

Kites have 2 pairs of adjacent sides equal.

1 Draw a different kite in each grid.

2 Mark the equal sides and equal angles.

3 Check each kite for line symmetry. Rule the line of symmetry with a coloured pencil or pen.

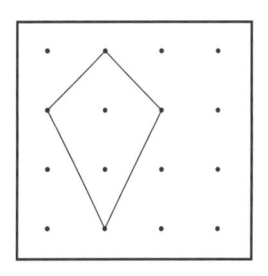

You need:
- ruler
- coloured pencil or pen

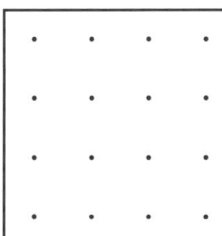

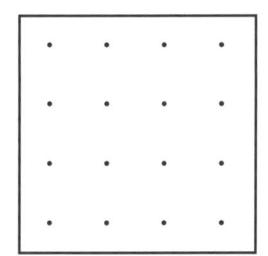

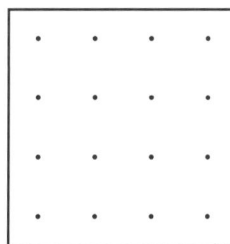

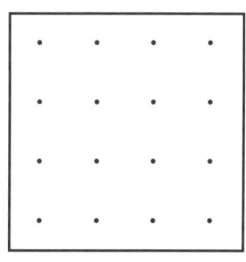

Follow these steps to make a kite from a sheet of A4 paper.

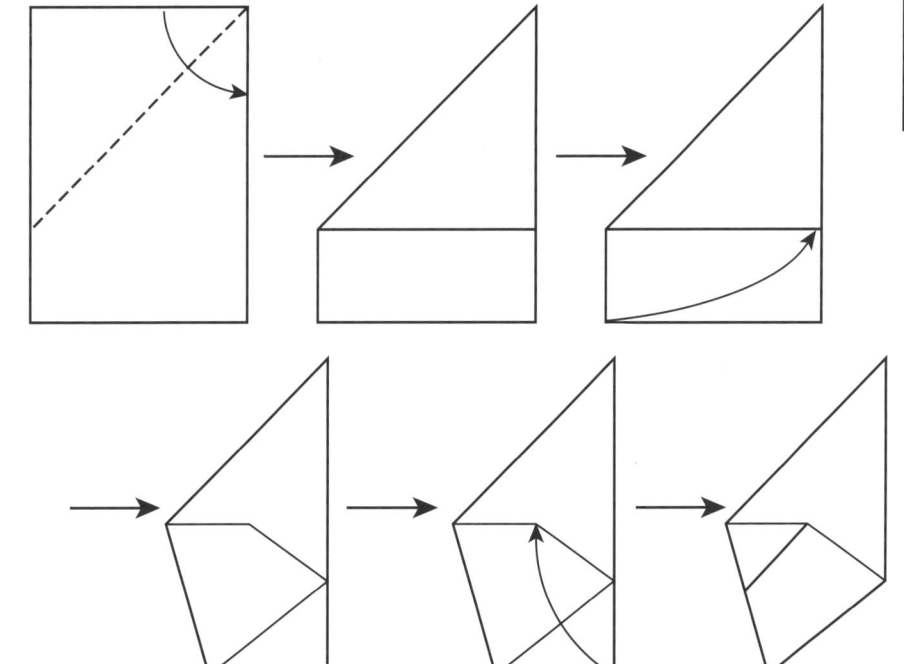

You need:
- A sheet of A4 paper

- Mark the equal sides and equal angles.
- Rule the line of symmetry.

© HarperCollinsPublishers Ltd 2007

Investigating midpoints

- Make and draw shapes accurately
- Describe and explain relationships

You need:
- ruler

Example

1.5 cm
3.0 cm

1. Find the midpoint on each of the two thicker sides of the triangle.

2. Use a ruler and sharp pencil to join the midpoints.

3. Measure to the nearest millimetre the length of:
   i the line connecting the midpoints
   ii the side of the triangle parallel to the midpoint line.

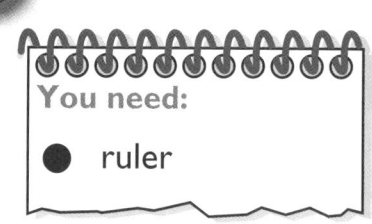

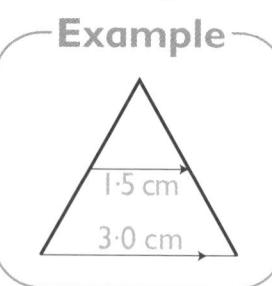

a

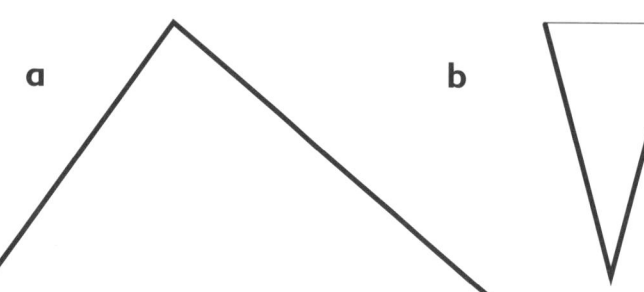

midpoint line = ☐ cm
parallel side = ☐ cm

b

midpoint line = ☐ cm
parallel side = ☐ cm

c

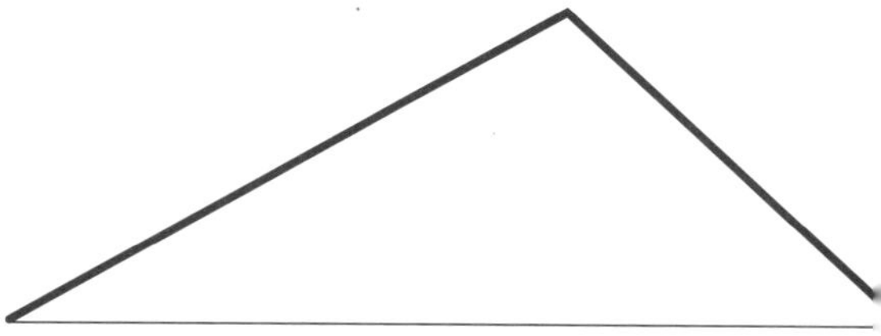

midpoint line = ☐ cm
parallel side = ☐ cm

d

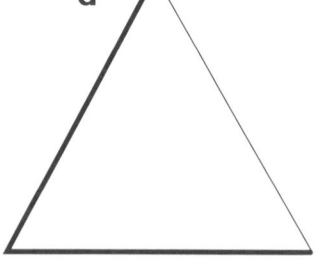

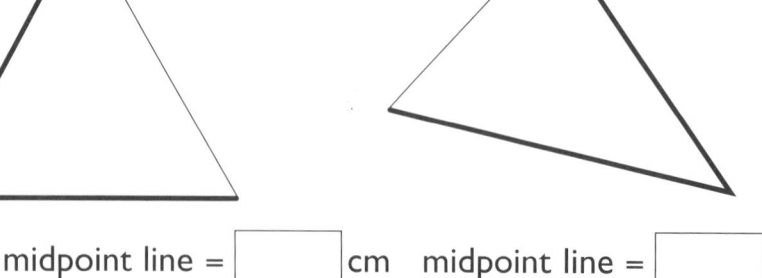

midpoint line = ☐ cm
parallel side = ☐ cm

e

midpoint line = ☐ cm
parallel side = ☐ cm

f

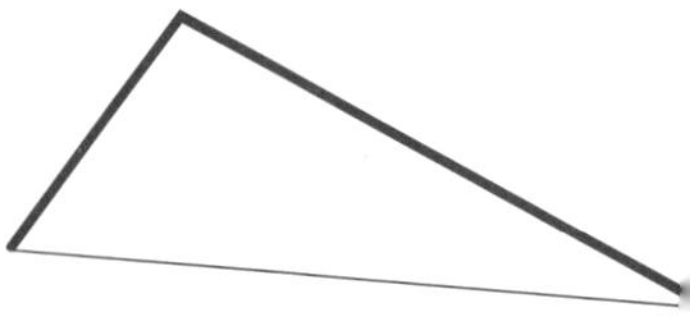

midpoint line = ☐ cm
parallel side = ☐ cm

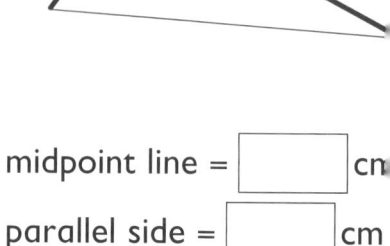

1. Find the midpoint of the third side and join all three midpoints.

2. Work out the perimeter of the large and small triangle in each example.
   Use the back of the sheet for your calculations.

3. Write what you notice.

© HarperCollinsPublishers Ltd 2007

All square measuring

- Read and interpret scales accurately

1. Find the length of one side of these squares to the nearest millimetre. Record your answers in the table below.

2. Work out the perimeter of each square.

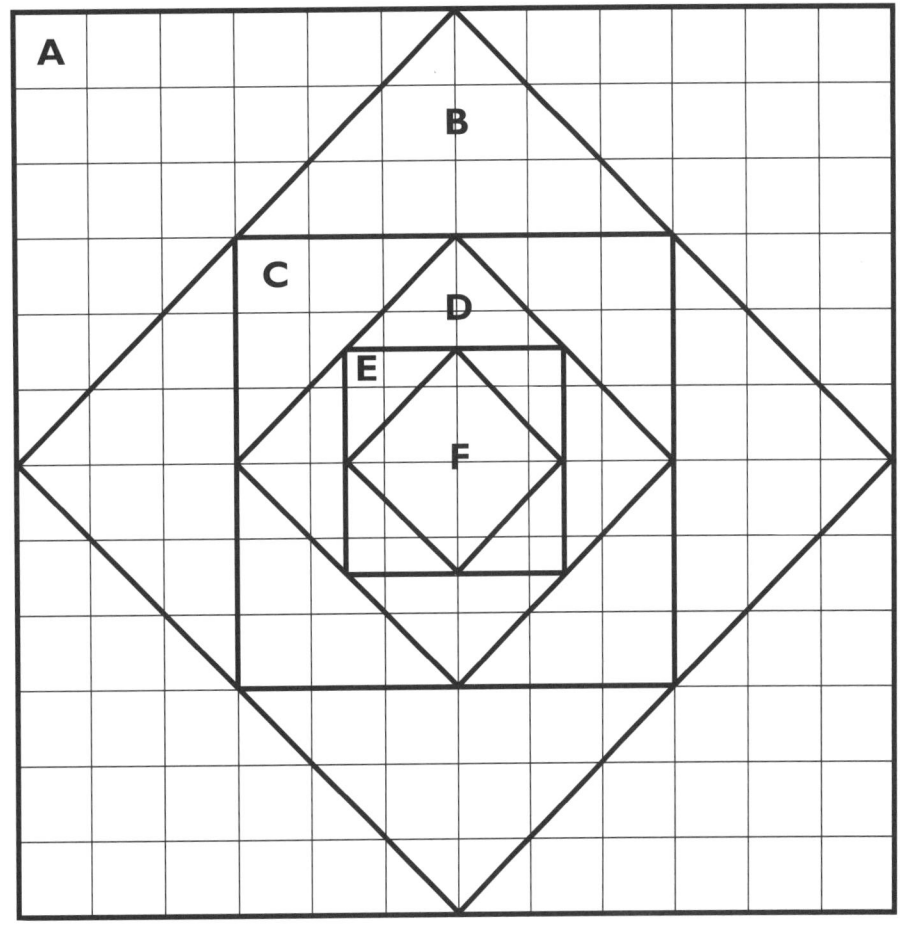

You need:
- ruler

Square	A	B	C	D	E	F
Length of side in cm	12	8·5				
Perimeter in cm	48					

1. Use the midpoints of square F to draw square G.

2. Complete these for square G: length of side in cm ☐ perimeter in cm ☐

3. Write what you notice about your results for squares A, C, E and G.

4. Suppose you went on drawing smaller and smaller squares. Find a way to work out the perimeter of square I.   Perimeter = ☐ cm

DIY ranges and modes

- Find the mode and range

1 a What is the smallest drill size for each brand? Drill Bits ☐ Smart Tools ☐

b What is the largest drill size for each brand? Drill Bits ☐ Smart Tools ☐

c Calculate the range for each brand. Drill Bits ☐ Smart Tools ☐

Drill Bits: 2 mm, 2 mm, 3 mm, 5 mm, 7 mm, 9 mm

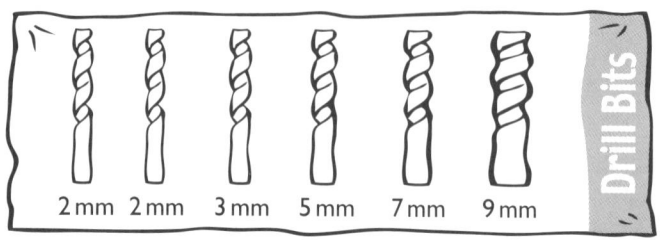

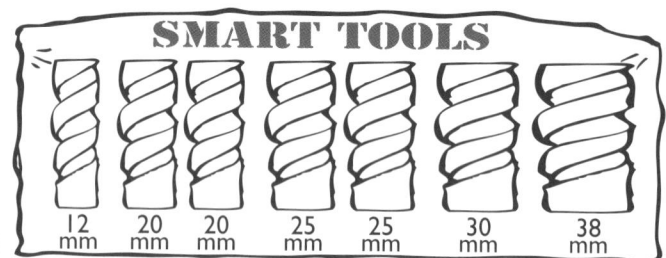

SMART TOOLS: 12 mm, 20 mm, 20 mm, 25 mm, 25 mm, 30 mm, 38 mm

2 a For each packet below, rearrange the screw sizes from smallest to largest.

b Calculate the range for each packet. Bowden's ☐ Perkins ☐

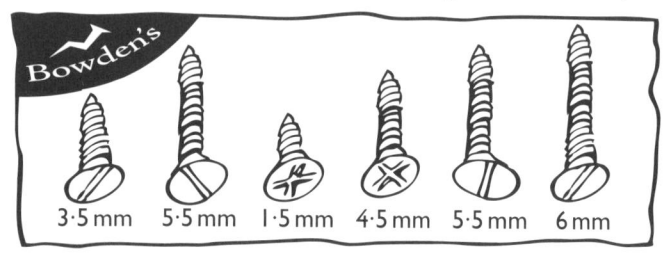

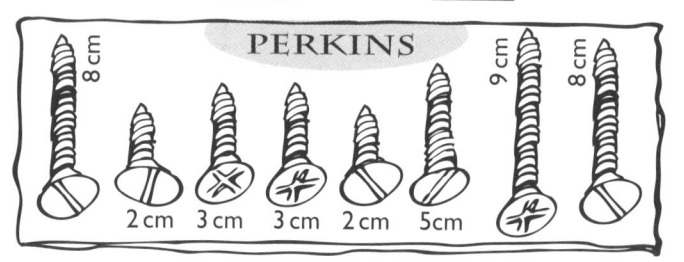

Bowden's: 3·5 mm, 5·5 mm, 1·5 mm, 4·5 mm, 5·5 mm, 6 mm

PERKINS: 8 cm, 2 cm, 3 cm, 3 cm, 2 cm, 5 cm, 9 cm, 8 cm

3 Find the modes for questions 1 and 2.

Drill Bits ☐ Smart Tools ☐ Bowden's ☐ Perkins ☐

1 Find the mode and range of staple sizes for each shop.

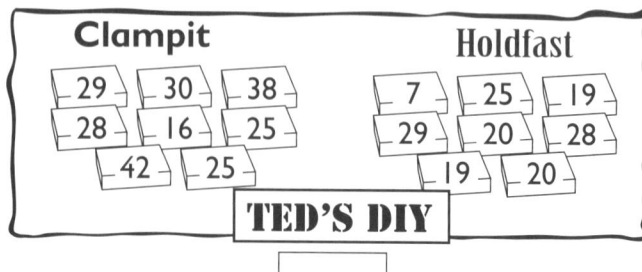

Clampit: 29, 30, 38, 28, 16, 25, 42, 25

Holdfast: 7, 25, 19, 29, 20, 28, 19, 20

TED'S DIY

Mode ☐
Range ☐

Gripper / Toolkit: 16, 24, 32, 20, 4, 18, 14, 20, 26, 16, 24, 16, 30, 32, 16, 22

Mode ☐
Range ☐

2 Use your statistics to compare the shops.

On the back of this sheet, write down a set of 10 drill sizes that have a range of 9 mm and a mode of 6 mm.

Documentary bar chart

- Represent data in different ways and understand its meaning

William recorded the lengths in minutes of news and documentary TV programmes. Here are his results.

15	25	45	28	10	10	25	60	47	40
30	30	30	20	5	36	30	45	35	8
28	30	25	50	8	30	5	10	30	55
40	40	15	25	30	35	25	5	50	36

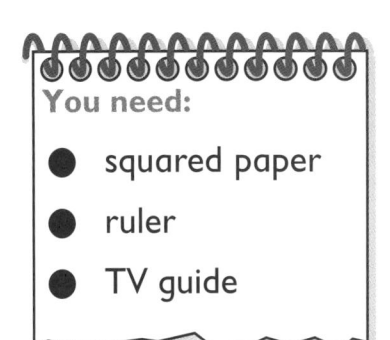

You need:
- squared paper
- ruler
- TV guide

Copy and complete the tally chart.

Length (minutes)	Tally	Total
1–10		
11–20		
21–30		
31–40		
41–50		
51–60		

1. Complete the bar chart.
2. a Which class has the most recorded programmes?

   b How many programmes last 20 minutes or less?

   c How many programmes last longer than 40 minutes?

Length of news and documentary TV programmes

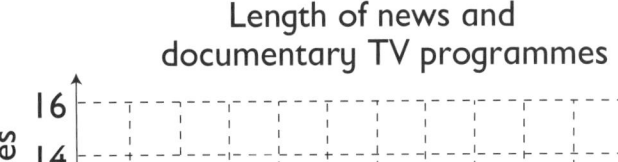

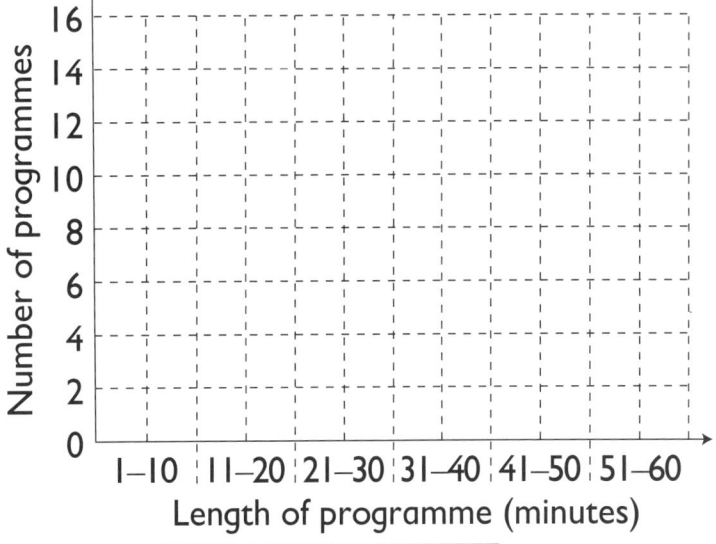

(Bar chart: Number of programmes (0–16) vs Length of programme (minutes): 1–10, 11–20, 21–30, 31–40, 41–50, 51–60)

1. What is the range of programme lengths?
2. How many programmes lasted between 21 and 50 minutes?
3. Use a TV guide to make your own table. Write your results on the back of this sheet. Compare the results.

Time for homework

- Find the mode and range
- Represent data in different ways and understand its meaning

■ Sara recorded the time, in minutes, she spent on her maths homework during Autumn term.

| 30 | 34 | 42 | 18 | 32 | 20 | 51 | 20 | 40 | 28 | 60 | 36 |
| 26 | 22 | 32 | 54 | 19 | 35 | 33 | 30 | 42 | 19 | 39 |

1. Find the range of homework times.
2. Complete the tally chart.

Time (minutes)	Tally	Total
10–19		
20–29		

3. Complete the bar chart.

Sara's homework times
(y-axis: Pieces of homework, 0–10; x-axis: 0–9, 10–19, 20–29, 30–39, 40–49, 50–59)

You need:
- squared paper

4. What does the tallest bar show?

● Bradley is in the same class as Sara, and recorded his homework times for Autumn.

| 48 | 20 | 34 | 22 | 27 | 10 | 31 | 30 | 18 | 18 | 51 | 20 |
| 23 | 22 | 25 | 39 | 27 | 40 | 28 | 9 | 35 | 16 | 26 |

1. Compare his range of times with Sara.
2. Make a tally chart for the data.
3. Draw a bar chart for the data.
4. Who do you think did their homework quicker overall? Explain using the bar chart

▲ Sara and Bradley recorded their first five homework times for the Spring term.

| Sara | 32 | 28 | 51 | 16 | 24 |
| Bradley | 29 | 17 | 34 | 20 | 48 |

Does this extra data affect any of your answers to the above questions? Do not draw new bar charts.

Name _____ Date _____

Rocky Mountain motoring

- Measure and calculate using imperial units (miles and kilometres)

Y6 D1 L3
HCM 15

In Canada, distances between destinations are given in both miles and kilometres.

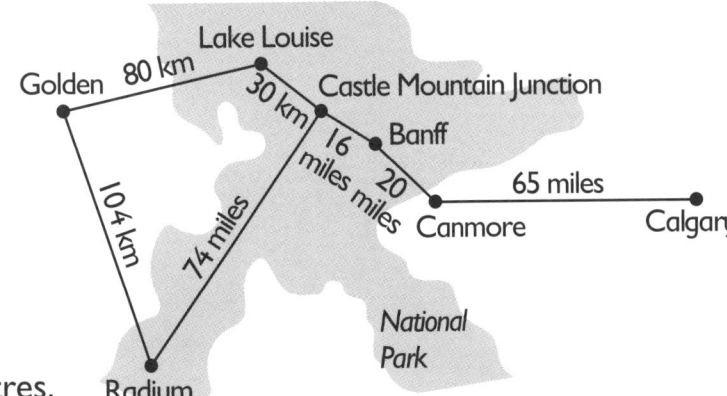

Map showing:
- Golden — 80 km — Lake Louise
- Lake Louise — 30 km — Castle Mountain Junction
- Castle Mountain Junction — 16 miles — (near Banff)
- Banff — 20 miles — Canmore
- Canmore — 65 miles — Calgary
- Golden — 104 km — Radium Hot Springs
- Radium Hot Springs — 74 miles — Castle Mountain Junction
- National Park

1 Find these distances in kilometres.
   Distance by road from Banff.

5 miles ≈ 8 kilometres

Town	miles	× 8	÷ 5	kilometres
Canmore	20	160	32	32
Calgary	85			
Lake Louise	35			
Radium Hot Springs	90			

2 Find these distances in miles.
   Distance by road from Golden.

8 kilometres ≈ 5 miles

Town	kilometres	÷ 8	× 5	miles
Lake Louise	80			
Radium Hot Springs	104			
Castle Mountain	110			
Banff	136			

You are staying at Lake Louise.
Work out the distance, in miles and kilometres for this round trip. Use the back of this sheet for all your working.

Lake Louise to Castle Mountain Junction
        to Radium Hot Springs
        to Golden
        back to Lake Louise

Answer

miles	kilometres

© HarperCollinsPublishers Ltd 2007

Collins New Primary Maths

Y6 D1
HCM

Name _____ Date _____

Finding perimeters

- Calculate the perimeter of simple compound shapes that can be split into rectangles

1 Find the perimeter of these shapes.

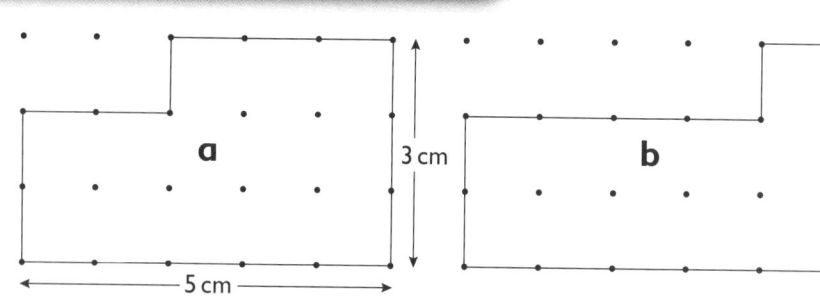

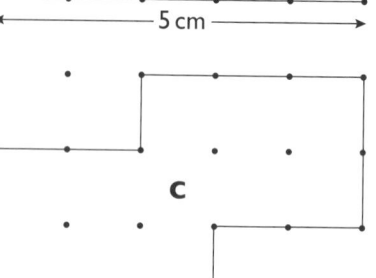

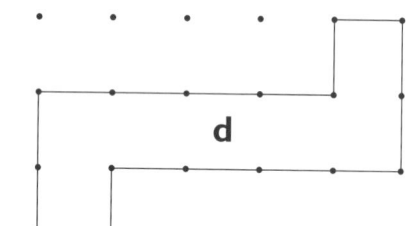

a P = ☐

b P = ☐

c P = ☐

d P = ☐

(shapes a, b, c, d shown on dot grids; shape a labelled with 3 cm and 5 cm)

2 Write what you notice.

☐

1 Work out the perimeter outlines of these shapes.

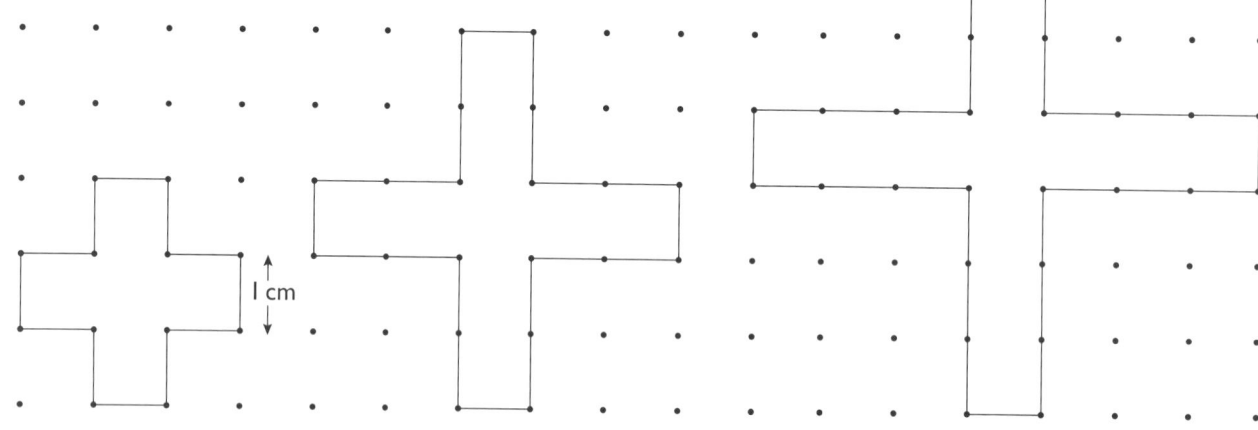

(shapes on dot grid, with 1 cm scale indicated)

2 Record your results in a table.

Shape (S)	1	2	3	4	5
Perimeter (P)					

3 Look for a pattern and complete the table.

1 Write a formula for the perimeter. P = ☐

2 Use the formula to find the perimeter outline for these shapes:

a shape 10   P = ☐

b shape 15   P = ☐

© HarperCollinsPublishers Ltd 2007

Y6 D1 L6
HCM 17

Name _____ Date _____

Area workout

- Calculate the area of a shape formed from rectangles

HINT
Use the dotted lines to help you.

1. Work out the area of these shapes in three different ways. Show all your working.

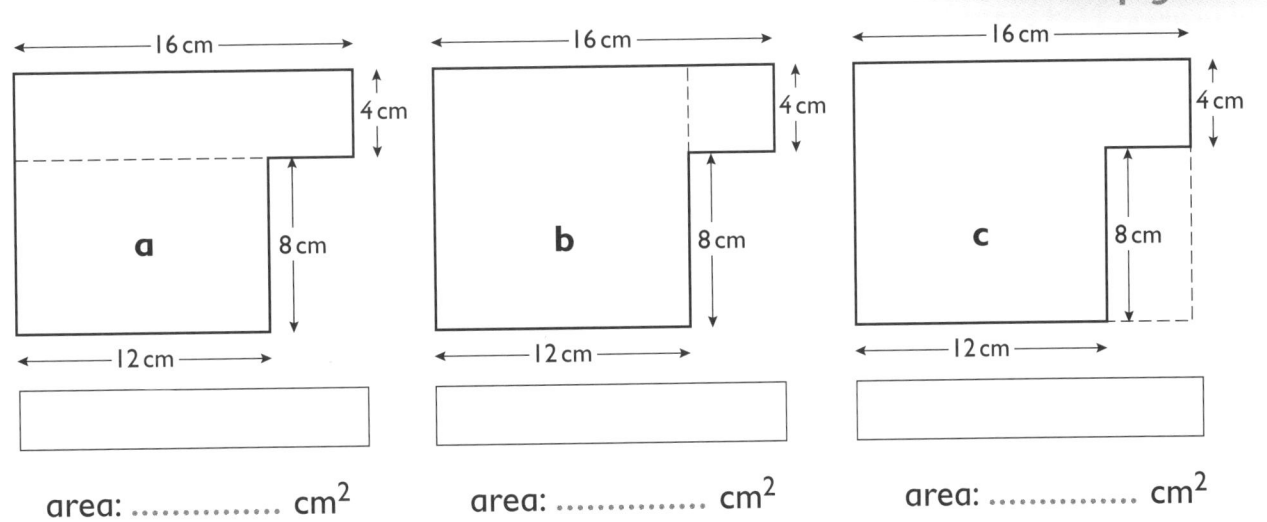

Shape a: 16 cm across the top, 4 cm down on the right, 8 cm on the right side, 12 cm along the bottom.

area: .... cm²

Shape b: 16 cm across the top, 4 cm down, 8 cm, 12 cm along the bottom.

area: .... cm²

Shape c: 16 cm across the top, 4 cm down, 8 cm, 12 cm along the bottom.

area: .... cm²

1. Work out the shaded area of these shapes.

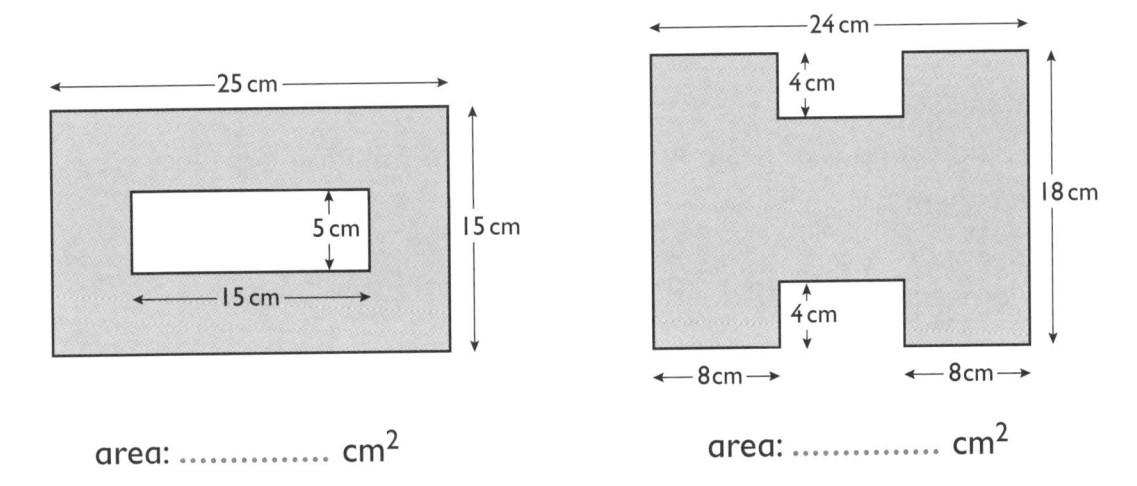

Shape 1: Rectangle 25 cm by 15 cm with a rectangular hole 15 cm by 5 cm.

area: .... cm²

Shape 2: H-shape, 24 cm wide, 18 cm tall, with 4 cm notches (8 cm wide) cut from top and bottom middle.

area: .... cm²

2. This table shows the area and the perimeter of various rectangles. Find the dimensions of these shapes.

rectangle	area	perimeter	length, l	breadth, b
a	42 cm²	26 cm	7 cm	cm
b	60 cm²	38 cm		
c	88 cm²	52 cm		
d	100 cm²	58 cm		
e	108 cm²	78 cm		

Remember
area = l × b
perimeter = 2 × (l + b)

© HarperCollinsPublishers Ltd 2007

Making calculations

- Use efficient written methods to add whole numbers and decimal numbers

■ Use these digits to write 5 addition calculations. Then work them out.

| 2 | 1 | 7 | 3 | 4 | 5 |

a 712 + 543 ───── 1255	**b**	**c**
d	**e**	**f**

● Use these digits and the decimal point to write 6 addition calculations. Then work them out.

| 8 | 3 | 0 | 7 | 6 | 3 | 9 | 5 | . |

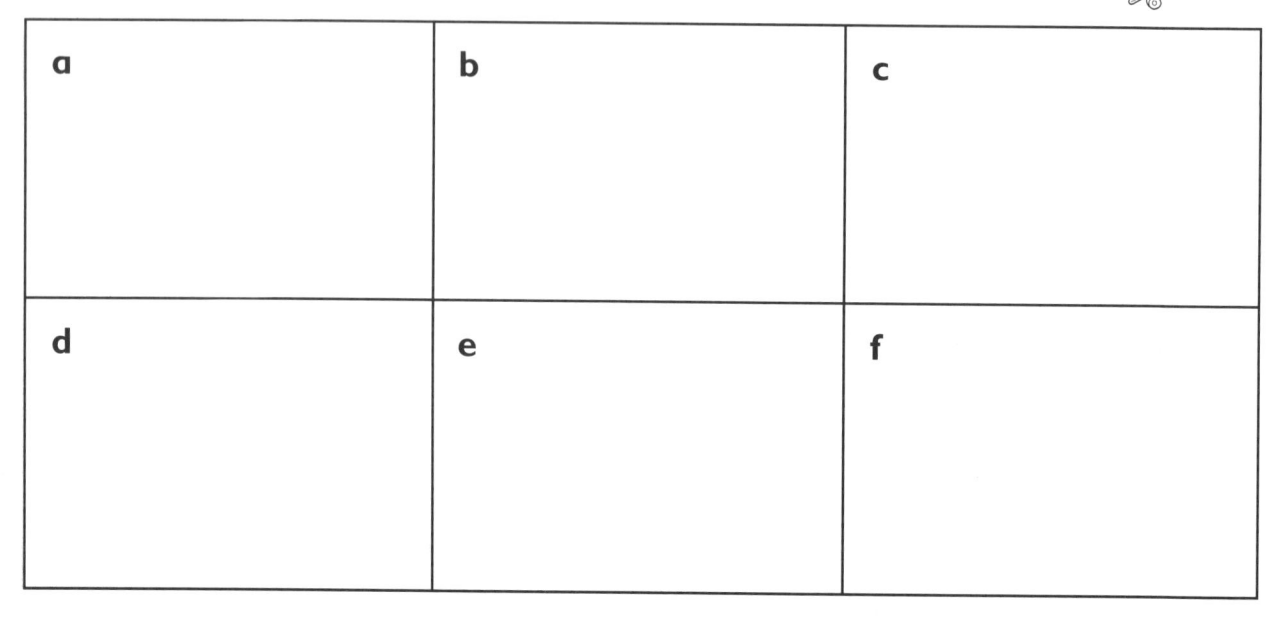

a	**b**	**c**
d	**e**	**f**

Name _____ Date _____

Y6 E1 L2
HCM 19

Colouring fractions

- Understand improper fractions and mixed numbers

Colour in the shapes. Describe each coloured fraction as an improper fraction and a mixed number.

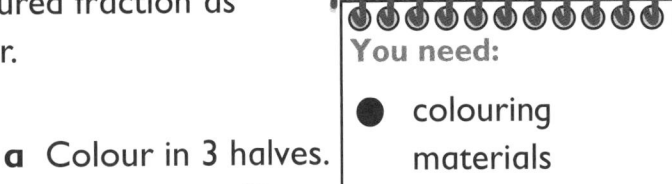

You need:
- colouring materials

Colour in 5 quarters.

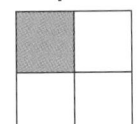

Improper fraction $\dfrac{5}{4}$

Mixed number $1\dfrac{1}{4}$

a Colour in 3 halves.

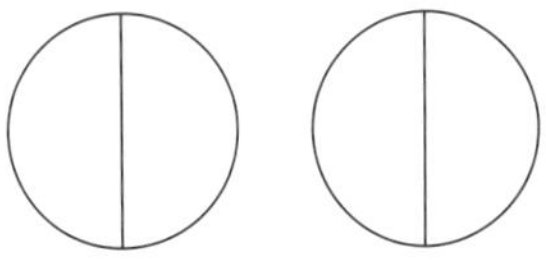

Improper fraction $\dfrac{\square}{2}$

Mixed number $\square\dfrac{\square}{2}$

b Colour in 7 sixths.

Improper fraction $\dfrac{\square}{6}$

Mixed number $\square\dfrac{\square}{6}$

c Colour in 5 thirds.

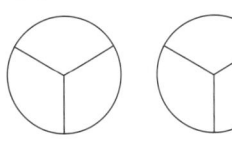

Improper fraction $\dfrac{\square}{3}$

Mixed number $\square\dfrac{\square}{3}$

d Colour in 8 fifths.

Improper fraction $\dfrac{\square}{5}$

Mixed number $\square\dfrac{\square}{5}$

e Colour in 10 sevenths.

Improper fraction $\dfrac{\square}{7}$

Mixed number $\square\dfrac{\square}{7}$

f Colour in 16 tenths.

Improper fraction $\dfrac{\square}{10}$

Mixed number $\square\dfrac{\square}{10}$

g Colour in 11 eighths.

Improper fraction $\dfrac{\square}{8}$

Mixed number $\square\dfrac{\square}{8}$

© HarperCollinsPublishers Ltd 2007

Sharing pizzas

- Relate fractions to division

◆ How much pizza would each person get if:
   a there were 6 friends?
   b there were 4 friends?
   c there were 5 friends?
   d there were 10 friends?

Write your answers as fractions.

● 1 How much pizza would each person get if:
   a there were 8 friends?
   b there were 10 friends?
   c there were 6 friends?
   d there were 12 friends?

2 Now write all of your answers as a different but equivalent fraction.

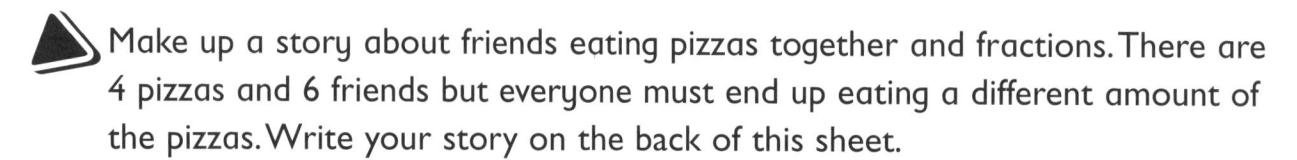

▲ Make up a story about friends eating pizzas together and fractions. There are 4 pizzas and 6 friends but everyone must end up eating a different amount of the pizzas. Write your story on the back of this sheet.

Y6 E1 L5
HCM 21

Percentages at home

- Understand percentage as the number of parts in every 100

■ Find three items of clothing at home. Look at the labels.

● Find six items of clothing at home. Look at the labels.

Write what the items are made of on the labels below. Convert any you can to fractions.

▲ Round all the percentages to the nearest multiple of ten and then convert them to fractions. Write your answers on the back of this sheet.

© HarperCollinsPublishers Ltd 2007

Decimal subtraction

- Use efficient written methods to subtract decimals

1. Use these digits to make 6 calculations to work out. Make the numbers to two decimal places.

Then complete each calculation.

7 4 3 1 2 8

Example
$$314.72$$
$$-\ 48.31$$

1.
2.
3.
4.
5.
6.

2. Use these digits to make 6 calculations to work out. Make the numbers in each calculation with a different number of decimal places.

Then complete each calculation.

9 7 6 4 3 1 0

Example
$$76.139$$
$$-\ 40.7$$

1.
2.
3.
4.
5.
6.

Y6 E1 L HCM

© HarperCollins Publishers Ltd 2007

Y6 E1 L11
HCM 23

Multiplication methods

- Approximate first. Use efficient written methods to multiply whole numbers

Approximate the answer to each calculation.

Example
317 × 28 ≈ 300 × 30 = 9000

a 227 × 12 ≈
b 325 × 15 ≈
c 219 × 23 ≈
d 268 × 13 ≈
e 176 × 18 ≈
f 324 × 25 ≈
g 249 × 38 ≈
h 286 × 27 ≈

For calculations **c** to **h** above, use the compact method to work out the answer.

c 219
   × 23
   ─────
   ____ × 20
   ____ × 3
   ─────
   ____

d 268
   × 13

e 176
   × 18

f

g

h

Name _____ Date _____

Design a snowman

- Solve a problem and check that the answer is correct

Imagine you have just built a snowman.
Now add these finishing touches.

Choose one hat.

top hat    bowler    western

Choose one nose.

carrot    parsnip    pine cone

Choose one extra item.

scarf    pipe    broom

Use the back of this sheet for all your diagrams.

You can use a tree diagram   or   list the different arrangements

top hat + carrot + scarf or T + C + S

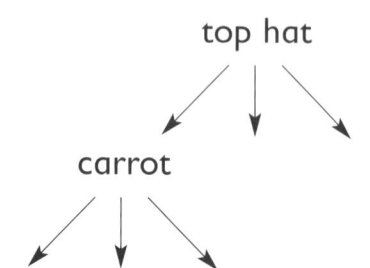

top hat

carrot

How many different snowmen can you make all wearing a top hat?

1 Find how many different snowmen you can make altogether.
2 What if you added sunglasses to the extra item choice?

Name _____ Date _____

Y6 A2 L1
HCM 25

Decimals decimals

• Order decimal numbers with up to three places

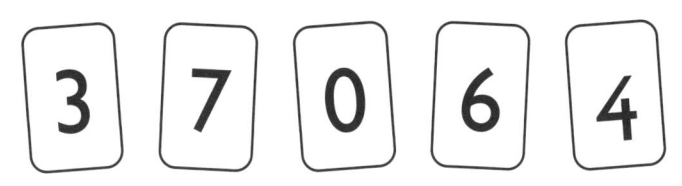

[3] [7] [0] [6] [4]

■ Use the digit cards to make 10 decimal numbers to two places, then write the decimal number that comes next.

▲ Use the digit cards to make 10 decimal numbers to three places, then write the decimal number that comes next.

▲ Use the digit cards to make 10 decimal numbers to three places and with the same whole number. Write your numbers on the back of this sheet from largest to smallest.

What is the largest decimal number you have made?

What is the smallest decimal number you have made?

What is the difference between them?

© HarperCollinsPublishers Ltd 2007

Y6 A2
HCM

Name _____ Date _____

Money problems

- Solve multi-step problems

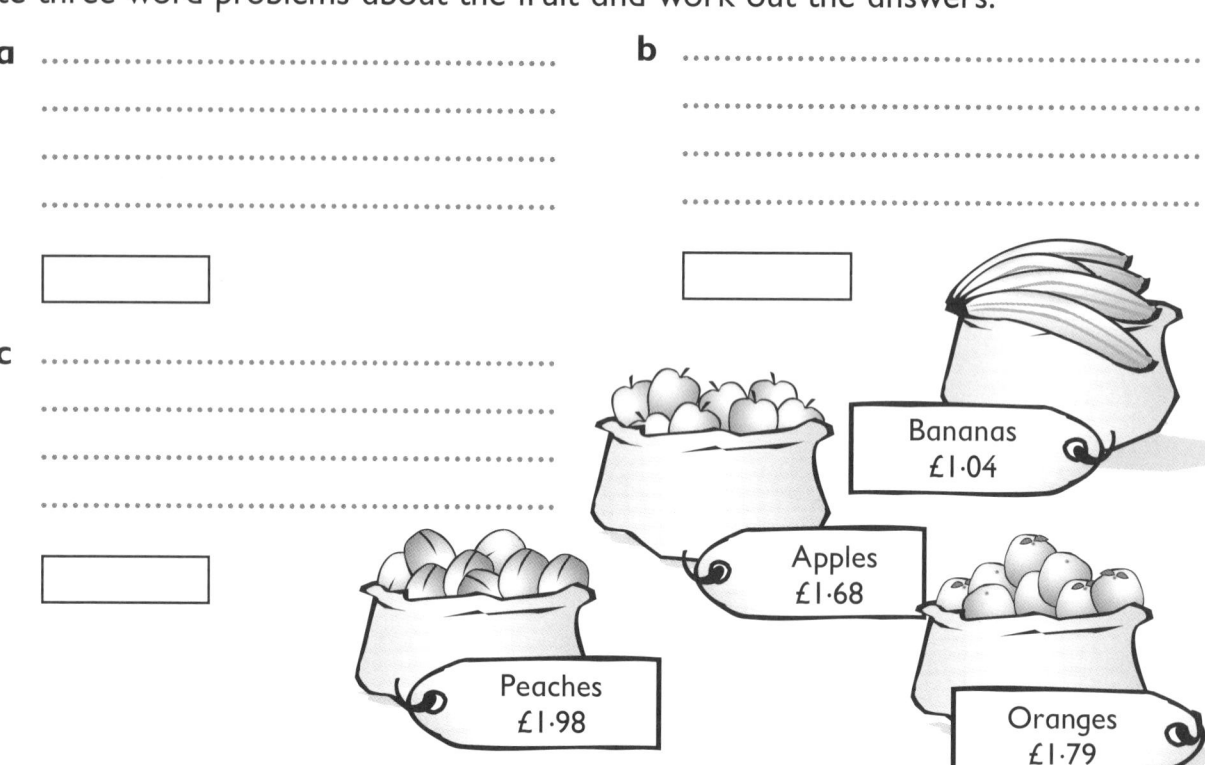

■ Write three word problems about the fruit and work out the answers.

a ....
....
....
....

b ....
....
....
....

c ....
....
....
....

Bananas £1·04
Apples £1·68
Peaches £1·98
Oranges £1·79

◆ Write three word problems about the food and work out the answers. Make them two-step problems.

a ....
....
....
....

b ....
....
....
....

c ....
....
....
....

Cheese £3·76 per kilogram
Pizza £4·96 for two
Doughnuts £2·50 for 10
Orange juice £1·44 for $1\frac{1}{2}$ litres

© HarperCollinsPublishers Ltd 2007

Multiplying mentally

- Multiply mentally with whole numbers and decimals

For each calculation, partition each number mentally. Write the answer to each calculation. Then write the final answer.

Example
46 × 5 = 200 + 30 = 230

(40 × 5) + (6 × 5)

1 a 36 × 4 = ...
   b 29 × 7 = ...
   c 56 × 5 = ...
   d 48 × 6 = ...
   e 79 × 8 = ...
   f 64 × 9 = ...

2 a 87 × 6 = ...
   b 48 × 8 = ...
   c 57 × 7 = ...
   d 68 × 9 = ...
   e 74 × 7 = ...
   f 56 × 6 = ...

Complete each multiplication grid.
Work out the answers in your head.

a

×	12	26	47	63
3	36			
4				
6				
12				

b

×	18	43	59	84
5				
7				
9				
4				

c

×	15	32	56	94
6				
8				
4				
7				

d

×	17	34	48	75
3				
9				
5				
8				

Y6 A2 L6
HCM 27

© HarperCollinsPublishers Ltd 2007

Family holiday

- Choose and use appropriate strategies to solve problems

PRICES (Return)
ADULTS: £260
Children (2–12 yrs): 25% discount
Infants (0–2 yrs): Free

PRICES
Per night
ADULTS: £25
Children: Half price

BREAKFAST
£3·50 per person per day

BREAKFAST AND DINNER
Add 25% to your room cost

Plan a holiday for your family.
Choose an appropriate method of calculating your answer:
- mental
- mental with jottings
- written (efficient method)

(Use the back of this sheet if you need more space for working out.)

Your family:
Number of adults ____
Number of children ____

a How much does it cost your family for the flights?	b What is the total cost of the hotel for your family if you stay for 7 nights?	c How much does it cost your family to have breakfast at the hotel each morning for 7 days?
d What is the additional cost to the hotel bill for your family to take breakfast and dinner?	e What is the difference in cost between breakfast and dinner, and breakfast only for your family for 1 week?	f What is the total cost for the week holiday if your family chooses breakfast and dinner while at the hotel?

© HarperCollinsPublishers Ltd 2007

Y6 B2 L2
HCM 29

Name _____ Date _____

Multiplication and division facts

- Use table facts to work out related facts

Work out the answers to these in your head.

a	120 ÷ 3 =	f	260 ÷ 13 =	k	450 ÷ 15 =
b	320 ÷ 8 =	g	320 ÷ 16 =	l	390 ÷ 13 =
c	480 ÷ 6 =	h	360 ÷ 60 =	m	420 ÷ 14 =
d	720 ÷ 9 =	i	420 ÷ 70 =	n	720 ÷ 12 =
e	450 ÷ 5 =	j	500 ÷ 25 =	o	880 ÷ 11 =

Complete the Crossnumber puzzle.
Find each answer and write the number in the puzzle.
Write only one digit in each box.

Across
b. 336 ÷ 12
e. 72 × 4
g. 26 × 23
i. 104 ÷ 4
j. 156 × 4
m. 14 × 27
o. 354 × 6

Down
a. 44 × 83
c. 744 ÷ 4
d. 936 ÷ 24
f. 2749 × 3
h. 258 ÷ 3
k. 936 ÷ 4
l. 637 ÷ 7
n. 8800 ÷ 100

Working

© HarperCollins Publishers Ltd 2007

Tests of divisibility (1)

- Know and apply simple tests of divisibility

Use your knowledge of divisibility tests.

Sort the numbers into the correct boxes. Some numbers belong in more than 1 box.

A number is divisible by:
2 if it is an even number and it ends in 0, 2, 4, 6 or 8
3 if the sum of its digits is divisible by 3
4 if the tens and units digits divide exactly by 4
5 if it ends in 0 or 5
6 if it is even and it is also divisible by 3
8 if half of it is divisible by 4 or if its last three digits are divisible by 8
9 if the sum of its digits are divisible by 9
10 if it ends in 0

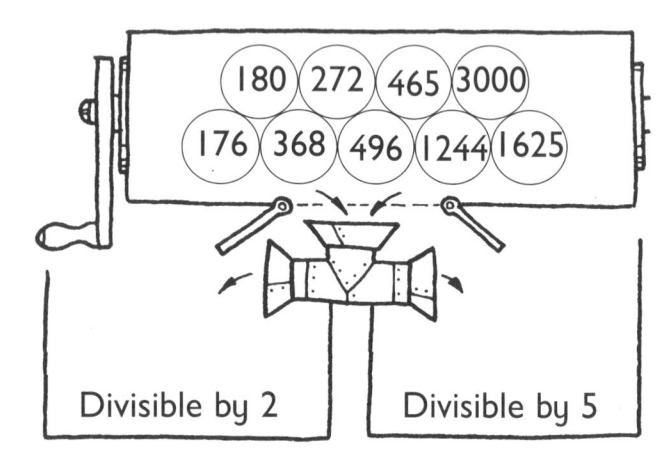

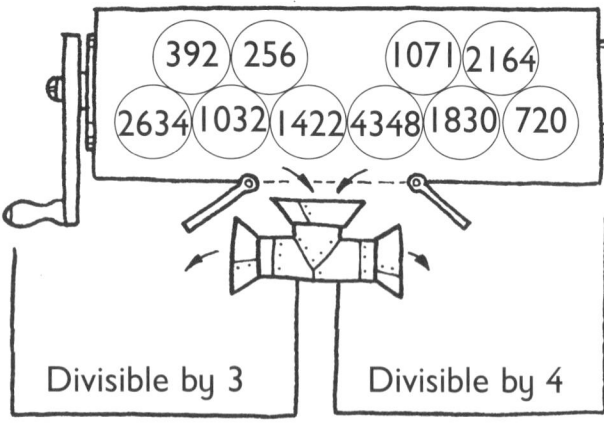

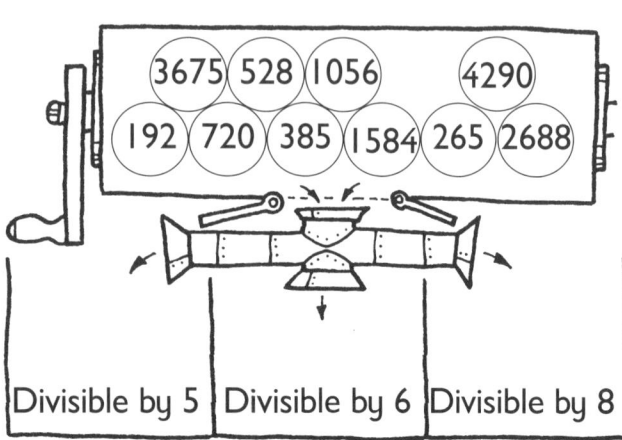

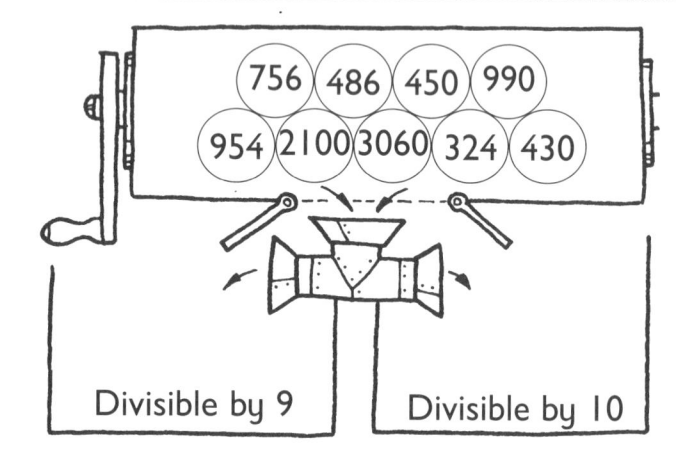

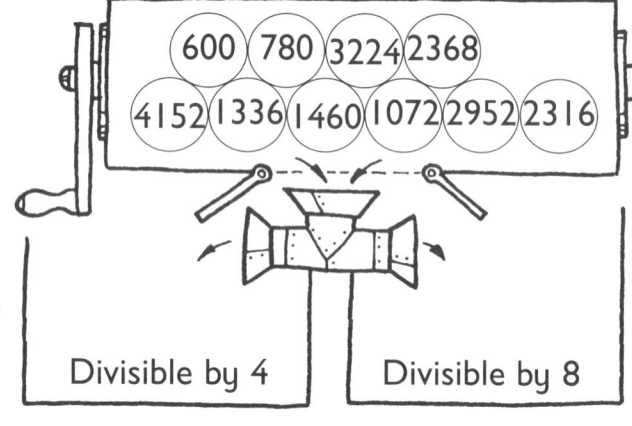

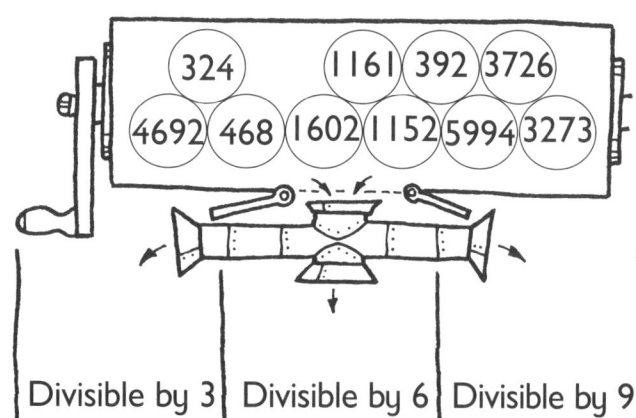

Numbers: 180, 272, 465, 3000, 176, 368, 496, 1244, 1625
- Divisible by 2
- Divisible by 5

Numbers: 756, 486, 450, 990, 954, 2100, 3060, 324, 430
- Divisible by 9
- Divisible by 10

Numbers: 392, 256, 1071, 2164, 2634, 1032, 1422, 4348, 1830, 720
- Divisible by 3
- Divisible by 4

Numbers: 600, 780, 3224, 2368, 4152, 1336, 1460, 1072, 2952, 2316
- Divisible by 4
- Divisible by 8

Numbers: 3675, 528, 1056, 4290, 192, 720, 385, 1584, 265, 2688
- Divisible by 5
- Divisible by 6
- Divisible by 8

Numbers: 324, 1161, 392, 3726, 4692, 468, 1602, 1152, 5994, 3273
- Divisible by 3
- Divisible by 6
- Divisible by 9

© HarperCollinsPublishers Ltd 2007

Y6 B2 L7
HCM 31

Using square numbers

- Know square numbers to 12 x 12

Write the missing numbers in the grid.
Do **not** write in the shaded boxes.

You need:
- calculator

+	1^2	2^2	3^2	4^2	5^2	6^2	7^2	8^2	9^2	10^2
1^2	2									
2^2		8								
3^2			18							
4^2				32						
5^2					50					
6^2						72				
7^2							98			
8^2								128		
9^2									162	
10^2										200

1   Work out the missing numbers to complete the equations.
    Use the grid to help you.

   a $\square^2 + \square^2 = 40$ b $\square^2 + \square^2 = 90$ c $\square^2 + \square^2 = 113$

   d $\square^2 + \square^2 = 29$ e $\square^2 + \square^2 = 106$ f $\square^2 + \square^2 = 73$

2   Sanjay listed the squares of 1, 11 and 111.
    He noticed a pattern.

    $1^2 = 1$
    $11^2 = 121$
    $111^2 = 12321$

   a   Complete the next two rows in the pattern.   .... = ...
       Check your answers with a calculator.   .... = ...

   b   Work out the answer to the 9th row.   .... = ...

© HarperCollins*Publishers* Ltd 2007

Name _____ Date _____

Prime numbers

- **Recognise prime numbers less than 100**

■ Circle the prime numbers in each set.

a	26	34	71	49	13	63	27
b	84	11	35	47	72	39	23
c	12	75	41	52	17	59	77
d	45	53	19	15	69	31	64
e	29	81	38	37	43	57	100

● Each of these numbers is the product of 2 prime numbers. What are they?

a 91 = ☐ × ☐
b 55 = ☐ × ☐
c 123 = ☐ × ☐
d 95 = ☐ × ☐
e 57 = ☐ × ☐
f 221 = ☐ × ☐
g 209 = ☐ × ☐
h 469 = ☐ × ☐
i 371 = ☐ × ☐
j 485 = ☐ × ☐

▲ Use the signs

☐ + ☐ ☐ − ☐ ☐ × ☐ ☐ ÷ ☐

to make each answer a prime number.

a 100 ☐ 60 ☐ 3 = ☐
b 4 ☐ 25 ☐ 3 = ☐
c 49 ☐ 7 ☐ 1 = ☐
d 24 ☐ 3 ☐ 11 = ☐
e 50 ☐ 40 ☐ 19 = ☐
f 20 ☐ 2 ☐ 12 ☐ 1 = ☐
g (25 ☐ 2) ☐ (3 ☐ 3) = ☐
h (15 ☐ 3 ☐ 1) ☐ 4 = ☐

Y6 B2 HCM

© HarperCollinsPublishers Ltd 2007

Y6 B2 L12
HCM 33

Name _____ Date _____

Co-ordinates of reflections

- Recognise where a shape will be after reflection.

1 a Reflect these half shapes in the mirror line.
   b Label each reflected vertex, join the points and complete the shape.

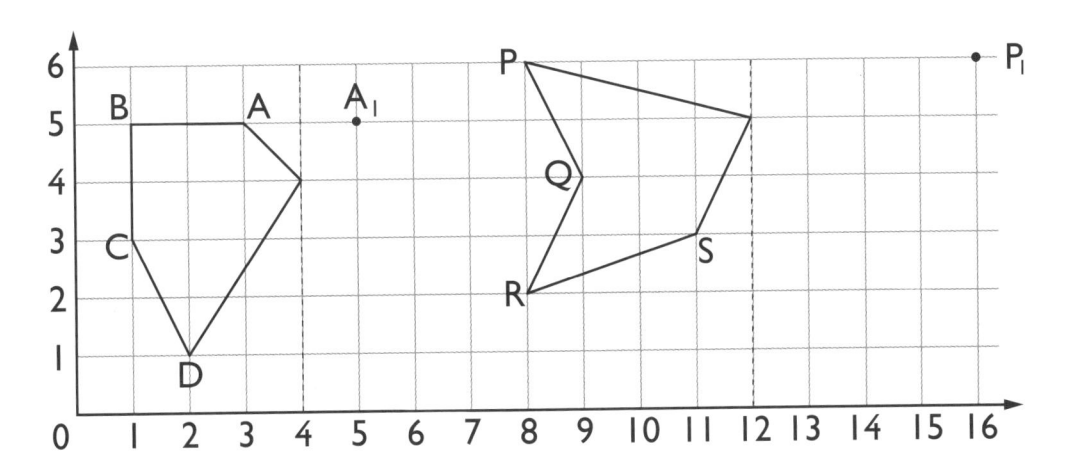

2 Complete these tables for the above shapes.

Left side		Right side	
vertex	co-ordinates	vertex	co-ordinates
A	(3, 5)	A₁	(5, 5)
B		B₁	
C		C₁	
D		D₁	

Left side		Right side	
vertex	co-ordinates	vertex	co-ordinates
P	(8, 6)	P₁	
Q		Q₁	
R		R₁	
S		S₁	

1 Reflect the shape in the mirror line, label the points and complete the table.

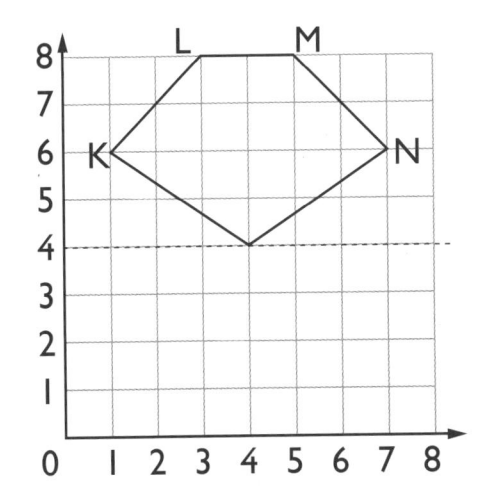

shape		reflection	
vertex	co-ordinates	vertex	co-ordinates
K		K₁	
L		L₁	
M		M₁	
N		N₁	

2 Write about the number patterns you notice on the back of this sheet.

© HarperCollinsPublishers Ltd 2007

Name _____ Date _____

Bisecting diagonals

- **Classify quadrilaterals**

1 Draw the 4 sides round these diagonals.
2 Mark with a small square the diagonals which intersect at right angles.

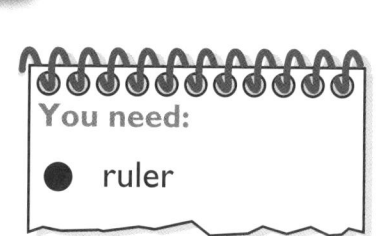

You need:
- ruler

a   b   c

d   e   f

g   h   i

1 List by letter and name, the shapes which have diagonals intersecting at right angles.

....

....

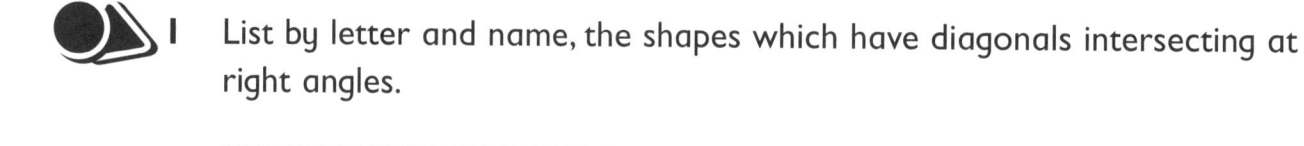

2 Name the shape which has diagonals of equal length, and give its letter.

....

Prehistoric weights

- Record units of measure (mass)

The scale shows the weight of dinosaurs in kilograms.

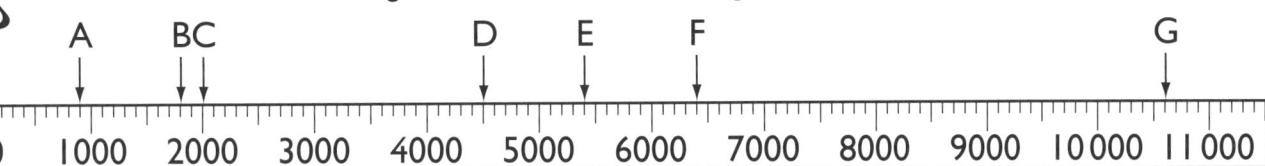

```
A       BC              D     E    F                              G
↓       ↓↓              ↓     ↓    ↓                              ↓
|0  1000  2000  3000  4000  5000  6000  7000  8000  9000  10000  11000|
```

1 Complete this table.

	Dinosaur	Weight in kilograms	Weight in tonnes to nearest $\frac{1}{10}$ tonne	Weight in tonnes to nearest tonne
A	megalosaurus			
B	stegosaurus	1800 kg	1·8 t	2 t
C	allosaurus			
D	iguanodon			
E	triceratops			
F	tyrannosaurus			
G	diplodocus			

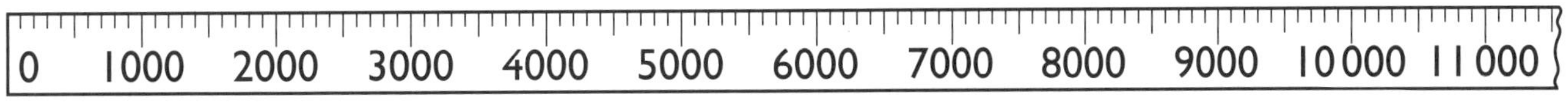

```
|0  1000  2000  3000  4000  5000  6000  7000  8000  9000  10000  11000|
```

2 Write the initial letter on the number line to show the position of these mammals.
E – elephant: 7 t   G – giraffe: 2 t   H – horse: 1·25 t   KW – killer whale: 9 t

Work out approximately how many:

a stegosaurus would balance an elephant ☐

b megalosaurus would balance a killer whale ☐

c giraffes would balance a diplodocus ☐

d horses would balance a tyrannosaurus ☐

Y6 C2 L1
HCM 35

© HarperCollinsPublishers Ltd 2007

Y6 C2
HCM

Name _____ Date _____

Mean and median

- Find the mean and median of a set of data

■ 1 Find the mean of each pair of numbers (add them up, divide by 2).
   a 2, 8   b 11, 25   c 30, 70
   Mean ☐   Mean ☐   Mean ☐

Remember
To find the mean, add the values and divide by the number of values.

2 Find the means of these numbers.
   a 5, 4, 6   b 9, 2, 10   c 20, 20, 20
   Mean ☐   Mean ☐   Mean ☐
   d 1, 3, 4, 4 e 3, 9, 0, 8 f 10, 20, 40, 10
   Mean ☐   Mean ☐   Mean ☐

● 1 Find the mean of these numbers.
   a 10, 15, 5   Mean ☐
   b 18, 8   Mean ☐
   c 5, 0, 0, 4, 11   Mean ☐

2 Find the mean of these weights.
   a 1 g, 2 g, 2 g, 5 g   Mean ☐
   b 50 g, 200 g, 200 g   Mean ☐

3 Find the mean of these numbers. If necessary, use a calculator.
   a 52, 70   Mean ☐
   b 28, 74, 32, 82   Mean ☐
   c 19, 99, 92   Mean ☐

You need:
- calculator (optional)

4 Find the mean of these prices. If necessary, use a calculator.
   a £9, £9, £3, £8, £7   Mean ☐
   b £1.42, £3.65, £8.25   Mean ☐

5 Find the mean weight of a cake.

Total weight: 220 g

▲ 1 Find the median of all the numbers in question 2 of the ■ section. Write your answers on the back of this sheet.

2 Find the medians of the numbers in each of questions 1, 2 and 3 of the ● section. Write your answers on the back of this sheet.

© HarperCollinsPublishers Ltd 2007

Beach time

- Find the median, mean and range
- Represent data in different ways and understand its meaning

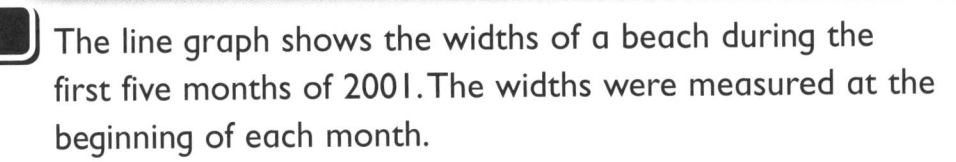

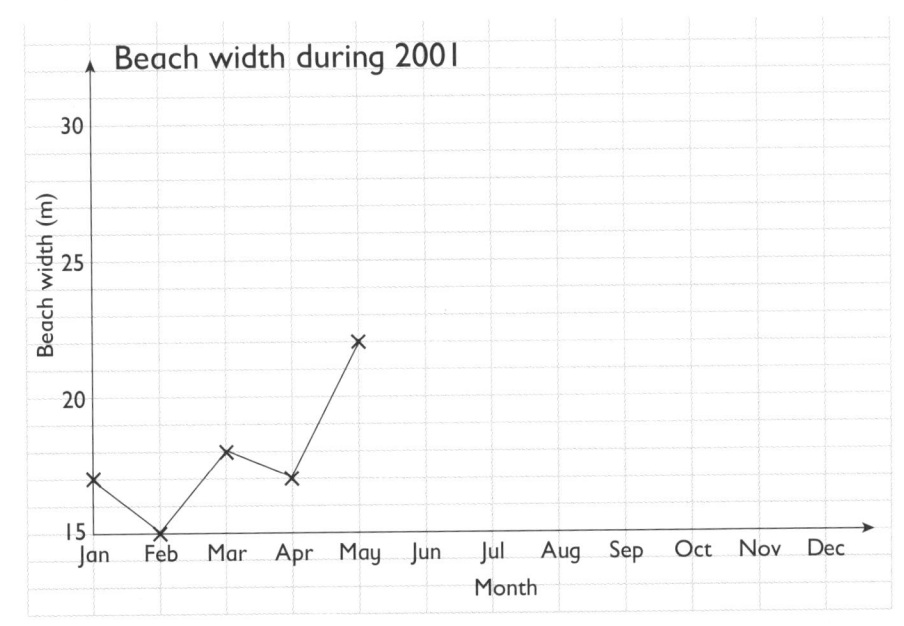

The line graph shows the widths of a beach during the first five months of 2001. The widths were measured at the beginning of each month.

You need:
- graph paper or 1 cm squared paper
- ruler
- calculator

Beach width during 2001

1 Write down the beach widths for the first five months.
2 Rearrange the widths from smallest to largest.
3 Find the median beach width.
4 Calculate the mean beach width.
5 Calculate the range of beach widths.

The table shows the beach widths for the rest of the year.

Month	Jun	Jul	Aug	Sep	Oct	Nov	Dec
Beach width (m)	23	20		28	20	19	16

1 Continue the line graph above.
2 Estimate the beach width at the beginning of August.
3 Calculate **a** the median ☐, **b** the mean ☐ and **c** the range ☐ of beach widths for June to December. (Don't include your value for August.)

Calculate **a** the median ☐, **b** the mean ☐ and **c** the range ☐ of beach widths for the whole year. (Don't include your value for August.)

Baked pie charts

- Answer questions using data from pie charts

The pie chart shows the different types of pies sold by a shop on Saturday afternoon.

They sold 60 pies.

1. Find the fraction sold for each type of pie.
2. Calculate the number sold of each type of pie.

Cherry
Apple
Mince
Banoffee

Total sold = 60

You need:
- calculator

Pie	Fraction	Number of pies
Cherry		
Apple		
Mince		
Banoffee		
	Total	

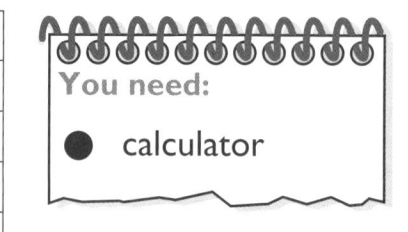

The pie chart shows the favourite pies of children of 64 children.

Complete the table.

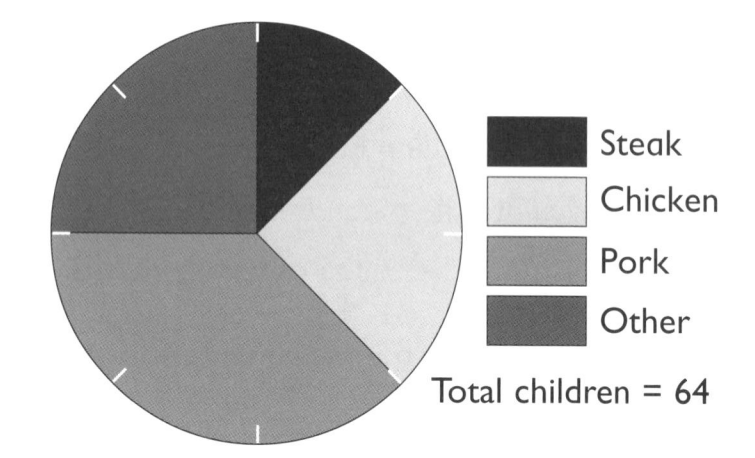

Steak
Chicken
Pork
Other

Total children = 64

Pie	Fraction	Number of children
Steak		
Chicken		
Pork		
Other		
	Total	

Conduct your own pie survey. Draw a pie chart to show the results.

Name _____ Date _____

Y6 D2
HCM

Two-step translations

- Use co-ordinates and translate shapes on grids

1 a Translate shape A 3 to the right, then 1 up to make shape B.
   b List the co-ordinates of shape B.
      (4, 2) (☐ , ☐) (☐ , ☐)

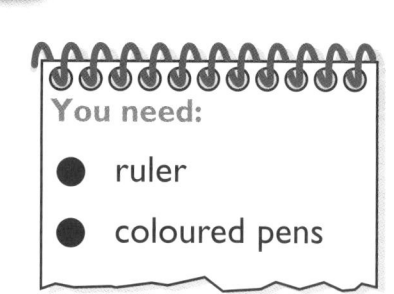

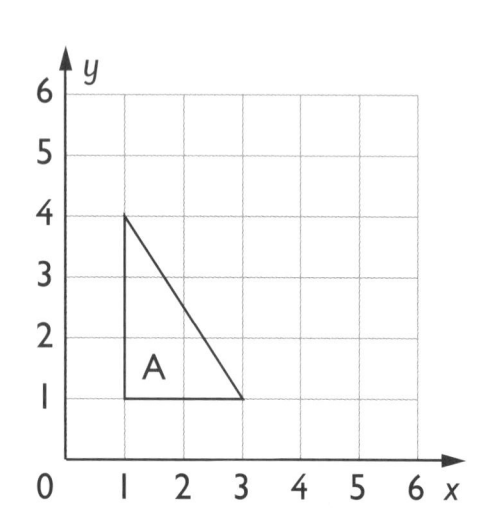

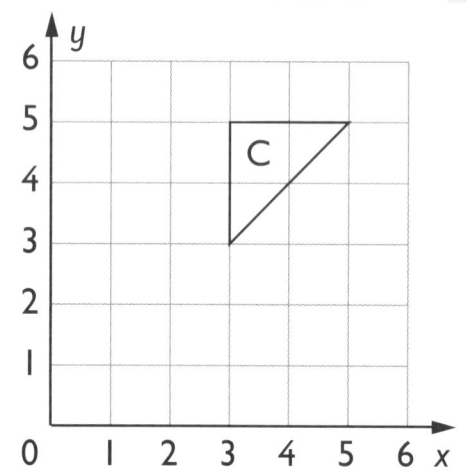

You need:
- ruler
- coloured pens

2 a Translate shape C 3 to the left, then 1 down to make shape D.
   b List the co-ordinates of shape D.
      (☐ , ☐) (☐ , ☐) (☐ , ☐)

1 Translate the shape 1 unit to the right, then 3 units down.

2 Repeat the translation.

3 Using the same rule, (1r, 3d) begin a translation of the same shape with its top left vertex at (4, 10). Continue to fill the grid and highlight the pattern in colour.

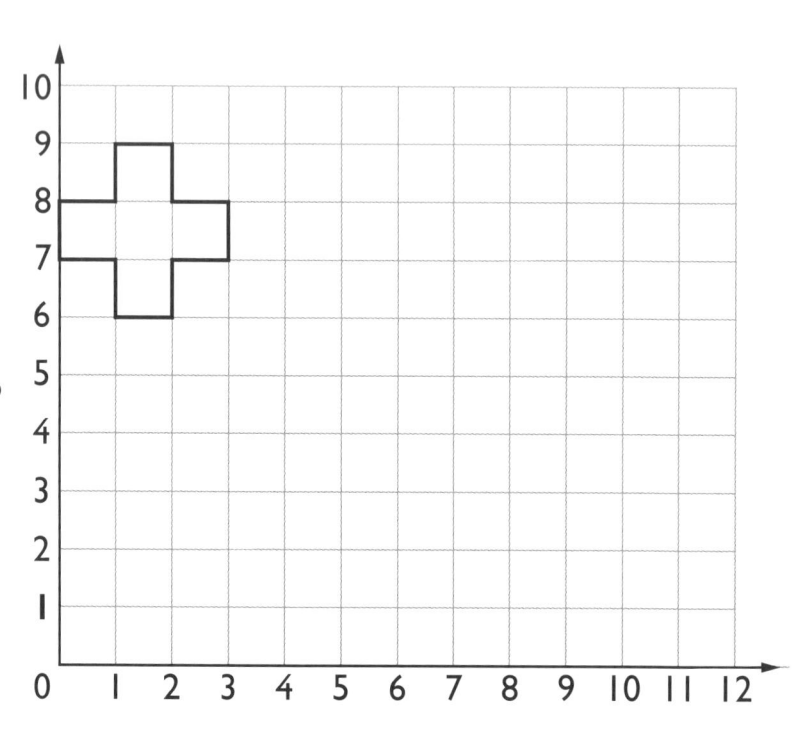

© HarperCollinsPublishers Ltd 2007

Measuring angles with set squares

- Use a protractor to measure and draw acute and obtuse angles to the nearest degree

You need:
- set squares (optional)

You can fit together the angles of 30°, 60° and 90° set squares to make different totals.

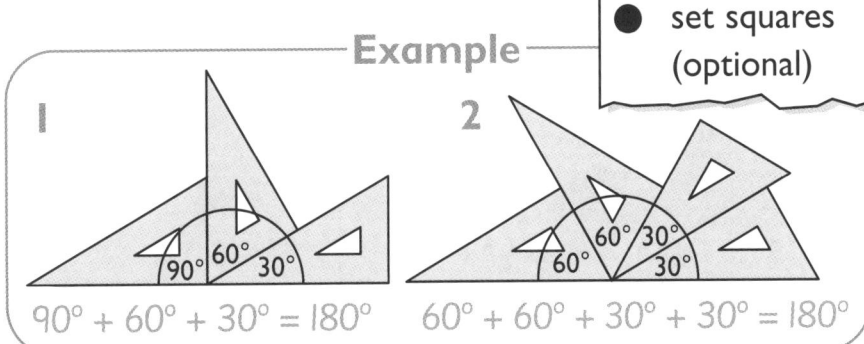

Example

1 90° + 60° + 30° = 180°

2 60° + 60° + 30° + 30° = 180°

1 Write the angles which fit together to make these totals:

Total	Angles used
180°	90° + 60° + 30°
210°	
240°	
270°	

2 Find two ways to fit 4 set squares together to make these reflex angles.

Reflex angle	1st way	2nd way
210°	90° + 60° + 30° + 30°	
240°		
270°		
300°		

Find three different ways to fit 5 set squares together to make these reflex angles.

Reflex angle	1st way	2nd way	3rd way
270°	(2 × 90°) + (3 × 30°)		
300°			
330°			

Inverse checking

- Use inverse operations to check results

Work out the following calculations using the inverse operation.

■

	Calculation	Inverse	Answer
1	☐ + 89 = 124		
2	☐ + 3·3 = 6·5		
3	142 + ☐ = 287		
4	186 + ☐ = 364		
5	5·9 + ☐ = 7·6		

●

	Calculation	Inverse	Answer
1	436 + ☐ = 805		
2	☐ − 247 = 596		
3	☐ − 742 = 1076		
4	394 + ☐ = 986		
5	572 − ☐ = 386		

▲

	Calculation	Inverse	Answer
1	372 + 167 − ☐ = 214		
2	685 − 254 + ☐ = 628		
3	3·7 + 6·8 − ☐ = 5·7		
4	843 − ☐ + 165 = 619		
5	☐ − 183 + 642 = 507		

Y6 E2 L1
HCM 43

Name _____ Date _____

Cancelling fractions

- Simplify fractions by cancelling common factors

■ Fill in the missing numbers.

a $\dfrac{1}{4} = \dfrac{\Box}{12}$ b $\dfrac{1}{6} = \dfrac{5}{\Box}$ c $\dfrac{1}{2} = \dfrac{\Box}{16}$

d $\dfrac{7}{21} = \dfrac{1}{\Box}$ e $\dfrac{4}{20} = \dfrac{1}{\Box}$ f $\dfrac{4}{32} = \dfrac{1}{\Box}$

g $\dfrac{3}{5} = \dfrac{\Box}{20}$ h $\dfrac{4}{7} = \dfrac{8}{\Box}$ i $\dfrac{2}{3} = \dfrac{\Box}{24}$

● Write these as fractions, then cancel them to their simplest form.

> 5 out of 20 = $\dfrac{5}{20} = \dfrac{1}{4}$

a 3 out of 6 = ☐ = ☐   b 2 out of 10 = ☐ = ☐

c 5 out of 25 = ☐ = ☐   d 4 out of 20 = ☐ = ☐

e 9 out 12 = ☐ = ☐   f 16 out of 20 = ☐ = ☐

▲ Cancel these fractions to their simplest form.

> $\dfrac{6}{18} = \dfrac{1}{3}$ (divide the top and bottom number by 6)

a $\dfrac{3}{9} =$ ☐   b $\dfrac{4}{20} =$ ☐   c $\dfrac{6}{12} =$ ☐

d $\dfrac{8}{12} =$ ☐   e $\dfrac{25}{30} =$ ☐   f $\dfrac{10}{16} =$ ☐

© HarperCollinsPublishers Ltd 2007

Y6 E2 HCM

Fraction wheels

- Find fractions of whole number quantities

Look at each number in the centre circle. Write the fractions of the number in the boxes.

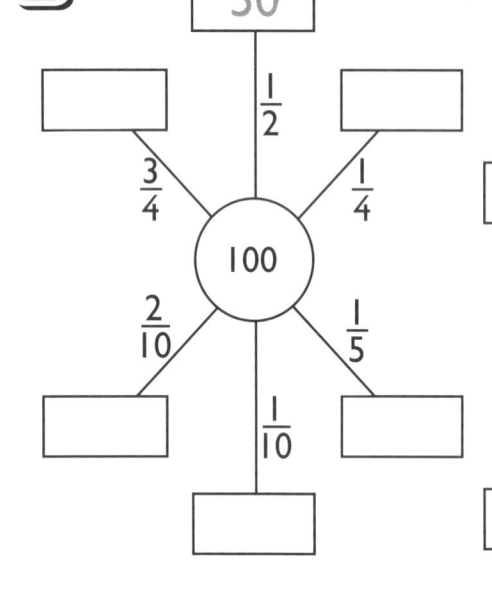

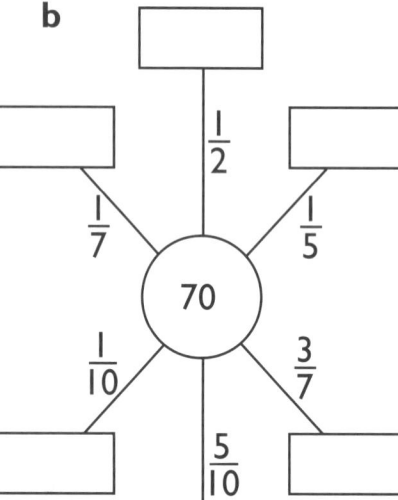

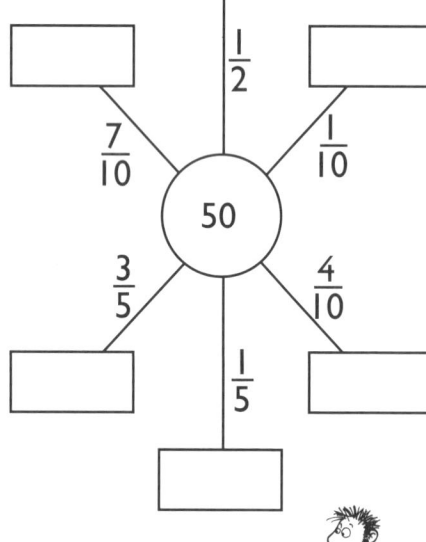

You need:
- calculator

1

a) Centre: 100. Fractions: $\frac{1}{2}$, $\frac{1}{4}$, $\frac{1}{5}$, $\frac{1}{10}$, $\frac{2}{10}$, $\frac{3}{4}$. Top box: 50.

b) Centre: 70. Fractions: $\frac{1}{2}$, $\frac{1}{5}$, $\frac{3}{7}$, $\frac{5}{10}$, $\frac{1}{10}$, $\frac{1}{7}$.

c) Centre: 50. Fractions: $\frac{1}{2}$, $\frac{1}{10}$, $\frac{4}{10}$, $\frac{1}{5}$, $\frac{3}{5}$, $\frac{7}{10}$.

2

a) Centre: £60. Fractions: $\frac{3}{4}$, $\frac{5}{6}$, $\frac{3}{10}$, $\frac{1}{15}$, $\frac{5}{12}$, $\frac{4}{6}$.

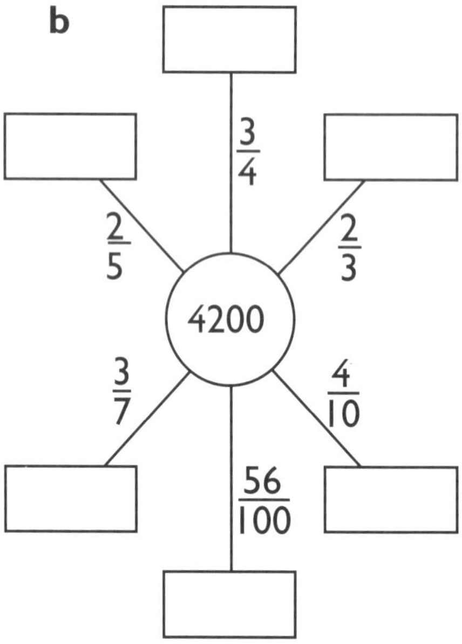

b) Centre: 4200. Fractions: $\frac{3}{4}$, $\frac{2}{3}$, $\frac{4}{10}$, $\frac{56}{100}$, $\frac{3}{7}$, $\frac{2}{5}$.

c) Centre: £84. Fractions: $\frac{2}{5}$, $\frac{3}{4}$, $\frac{7}{10}$, $\frac{5}{6}$, $\frac{45}{100}$, $\frac{3}{10}$.

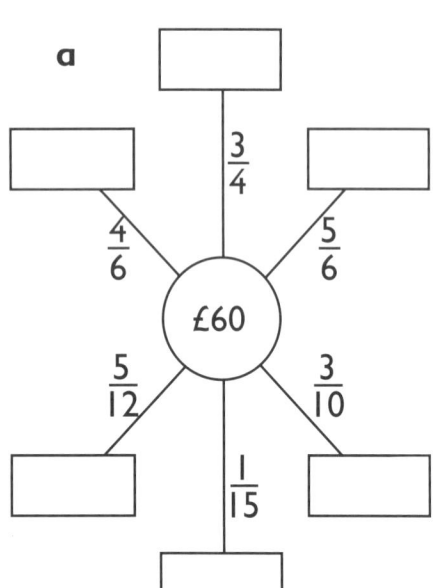

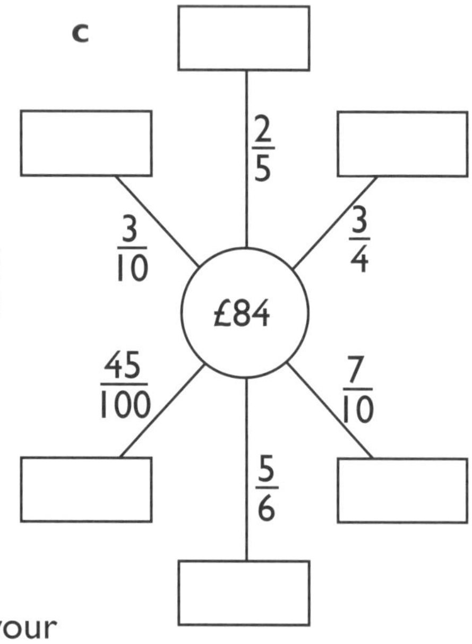

⚠ Explain how to work out fractions of amounts. Write your explanation on the back of this sheet.

© HarperCollinsPublishers Ltd 2007

Y6 E2 L4
HCM 45

Name _____ Date _____

Percentages of the day

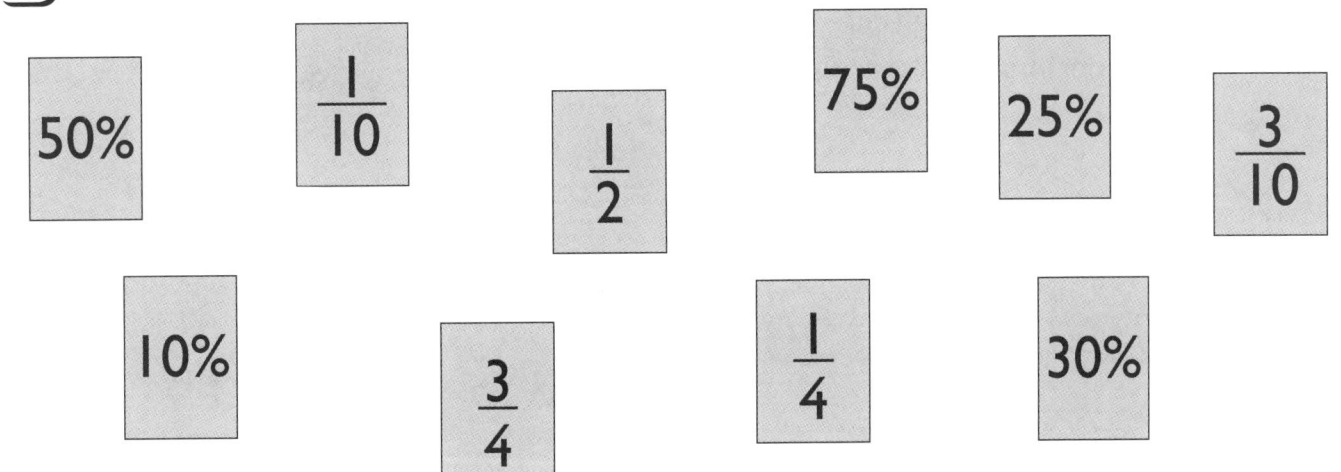

- Find percentages of whole number quantities

■ Draw a line to match each percentage to its fraction.

50% $\frac{1}{10}$ $\frac{1}{2}$ 75%   25% $\frac{3}{10}$

10% $\frac{3}{4}$ $\frac{1}{4}$ 30%

● Use this time line to represent part of your day. Put on a start time and then write in all the hours.

For each hour write down what you do. You can only write one thing for each hour.

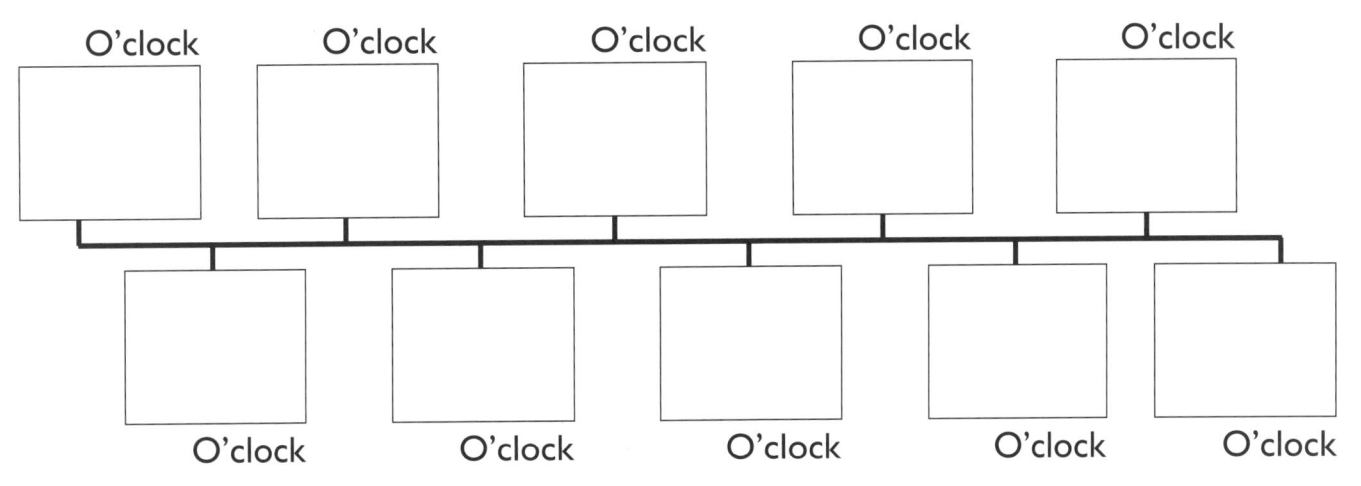

O'clock   O'clock   O'clock   O'clock   O'clock

O'clock   O'clock   O'clock   O'clock   O'clock

Now complete these statements.

I spend ☐ % of my time ☐

I spend ☐ % of my time ☐

I spend ☐ % of my time ☐

I spend ☐ % of my time ☐

▲ Before you complete the above timeline. Mark in half hours and then write in what you do in each half hour. Then complete the statements.

© HarperCollinsPublishers Ltd 2007

Y6 E2
HCM

Name _____ Date _____

Dominoes

- Find equivalent percentages, decimals and fractions

Cut out the dominoes. Make a domino ring by matching the equivalent decimals, percentages and fractions.

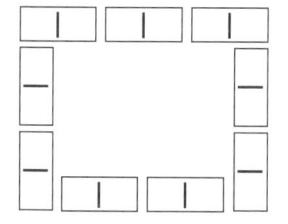

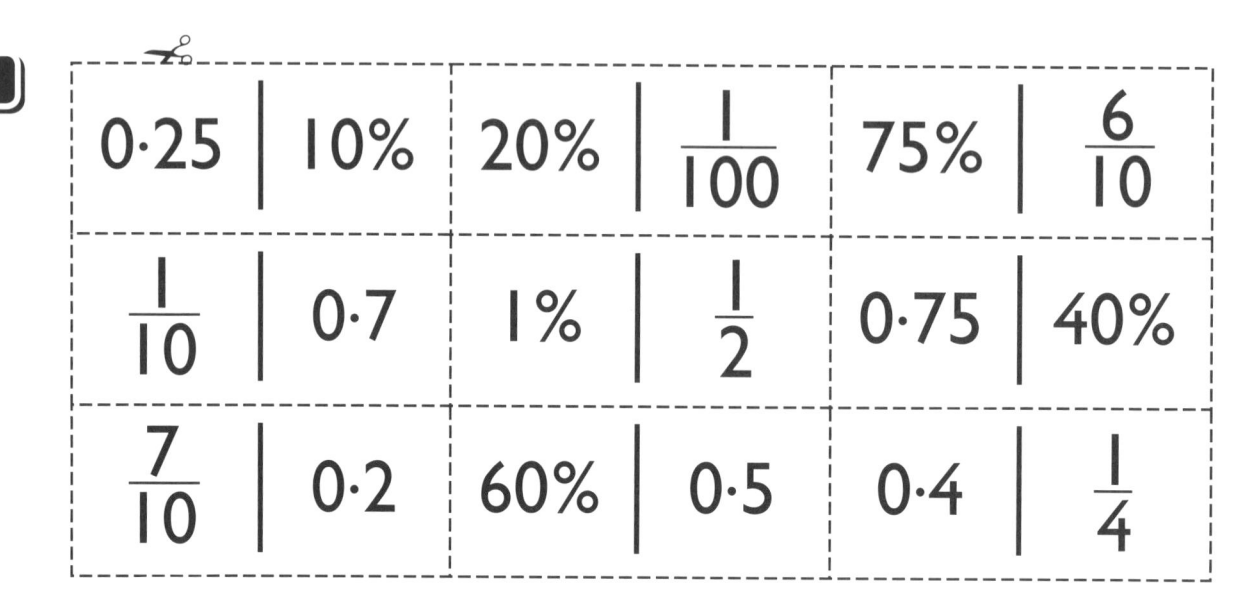

You need:
- scissors

■

0·25	10%	20%	$\frac{1}{100}$	75%	$\frac{6}{10}$
$\frac{1}{10}$	0·7	1%	$\frac{1}{2}$	0·75	40%
$\frac{7}{10}$	0·2	60%	0·5	0·4	$\frac{1}{4}$

●

$\frac{48}{100}$	$\frac{2}{5}$	0·4	$\frac{1}{3}$	$33\frac{1}{3}\%$	12·5%
$\frac{3}{4}$	$\frac{4}{5}$	0·03	80%	0·125	3%
0·6	75%	$37\frac{1}{2}\%$	$\frac{3}{5}$	48%	$\frac{3}{8}$

▲ Make up another domino ring using different equivalent fractions, decimals and percentages.

© HarperCollins Publishers Ltd 2007

Patterns, ratio and proportion

- Solve simple problems using ratio and proportion

■ Use the colours red and blue to make a pattern. What is the ratio of red to blue in your pattern?

You need:
- red and blue colouring materials

a ☐☐☐☐☐☐☐☐☐☐

The ratio is ☐ to every ☐

b ☐☐☐☐☐☐☐☐☐☐☐☐

The ratio is ☐ to every ☐

c ☐☐☐☐☐☐☐☐☐☐☐☐☐☐☐

The ratio is ☐ to every ☐

● a ☐☐☐☐☐☐☐☐☐☐☐☐☐☐☐☐☐☐☐☐

The ratio is ☐ to every ☐   The proportion is ☐ in every ☐

b ☐☐☐☐☐☐☐☐☐☐☐☐☐☐☐☐☐☐☐☐

The ratio is ☐ to every ☐   The proportion is ☐ in every ☐

c ☐☐☐☐☐☐☐☐☐☐☐☐☐☐☐☐☐☐☐☐

The ratio is ☐ to every ☐   The proportion is ☐ in every ☐

d ☐☐☐☐☐☐☐☐☐☐☐☐☐☐☐☐☐☐☐☐

The ratio is ☐ to every ☐   The proportion is ☐ in every ☐

▲ Explain what ratio and proportion are. How are they different? How are they related? Write your explanations on the back of this sheet.

Y6 E2 L
HCM

Name _____ Date _____

Mystic rose

- Tabulate systematically the information in a problem or puzzle

You make a mystic rose by joining the points which are equally spaced around the circle.

The 6 point mystic rose has 15 lines altogether. Use a ruler to join each dot to every other dot for this 16 point mystic rose. The 15 lines from point 1 are already drawn. Begin at point 2 and complete the design.

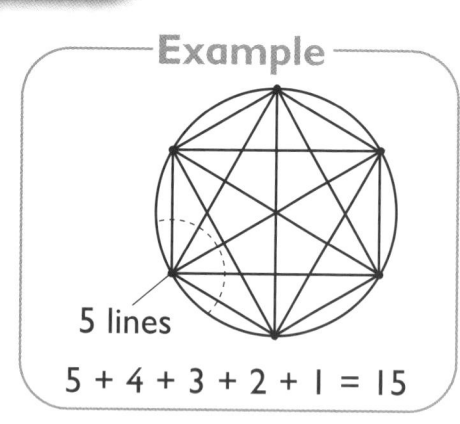

Example

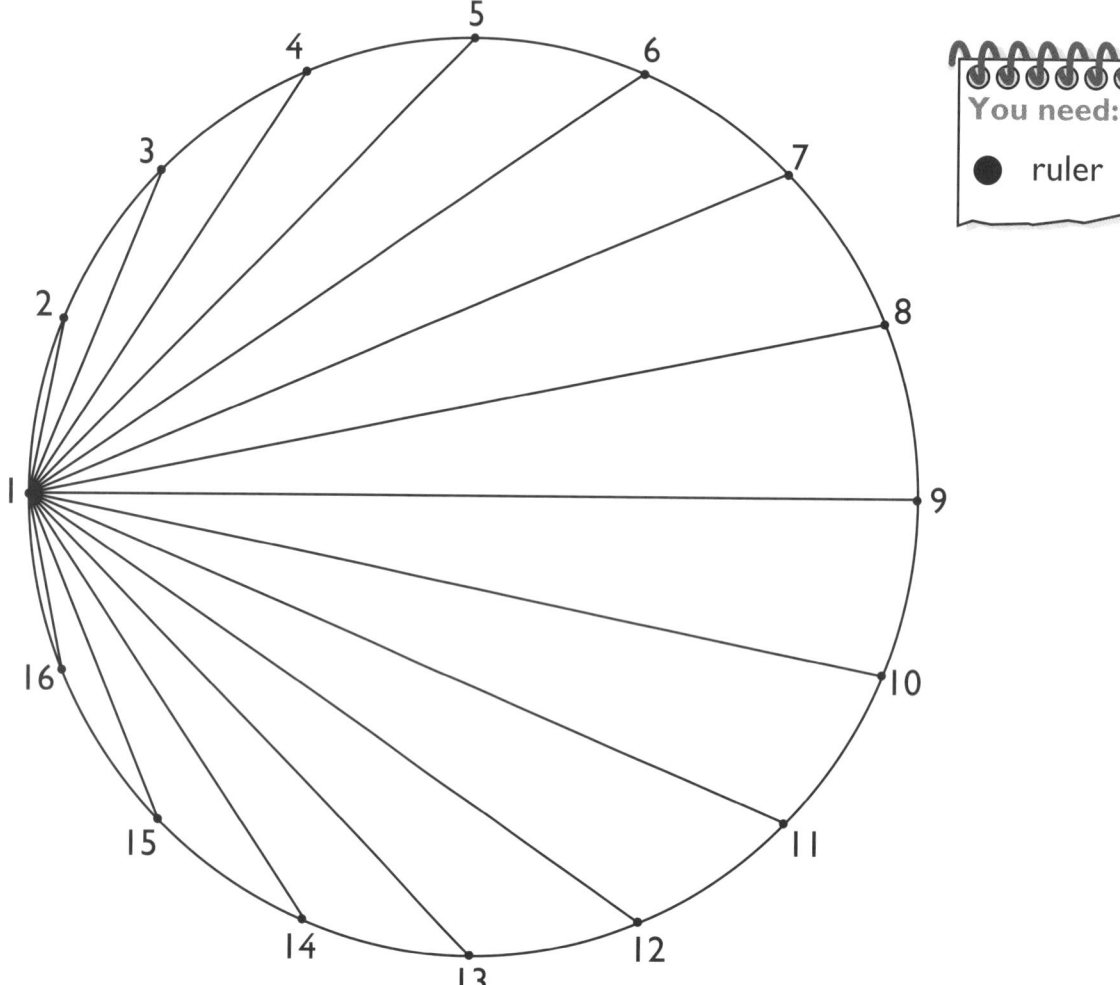

5 lines

$5 + 4 + 3 + 2 + 1 = 15$

You need:
- ruler

▲ Work out how many lines there are altogether in a 16 point mystic rose.

15 +   Answer: _____ lines

© HarperCollinsPublishers Ltd 2007

Order the decimals

- **Order decimals with up to three decimal places**

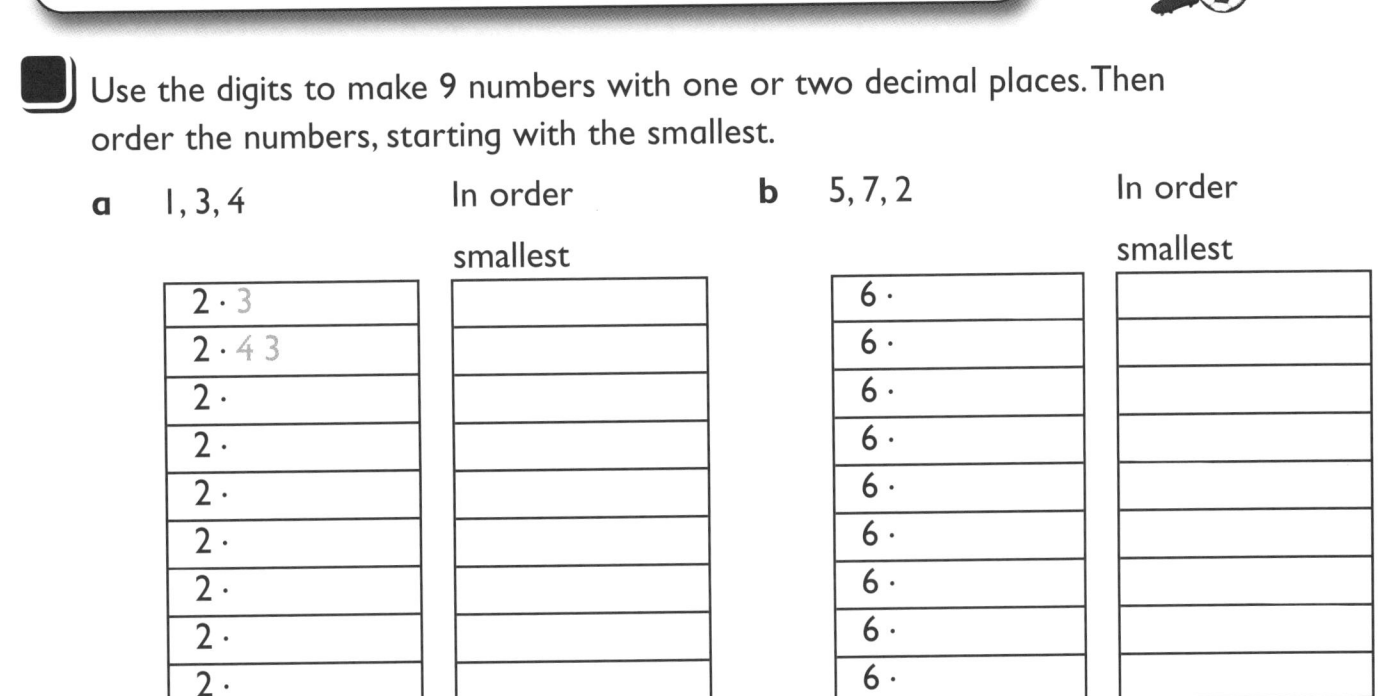

◼ Use the digits to make 9 numbers with one or two decimal places. Then order the numbers, starting with the smallest.

a 1, 3, 4

	In order
2·3	smallest
2·4 3	
2·	
2·	
2·	
2·	
2·	
2·	
2·	
	largest

b 5, 7, 2

	In order
6·	smallest
6·	
6·	
6·	
6·	
6·	
6·	
6·	
6·	
	largest

▶ Use the digits to make 14 numbers with one, two or three decimal places. Then order the numbers, starting with the smallest.

a 3, 4, 5

	In order
0·	smallest
0·	
0·	
0·	
0·	
0·	
0·	
0·	
0·	
0·	
0·	
0·	
0·	
0·	
	largest

b 9, 8, 2

	In order
6·	smallest
6·	
6·	
6·	
6·	
6·	
6·	
6·	
6·	
6·	
6·	
6·	
6·	
6·	
	largest

Y6 A3 L2
HCM 49

© HarperCollins*Publishers* Ltd 2007

Making calculations with decimals

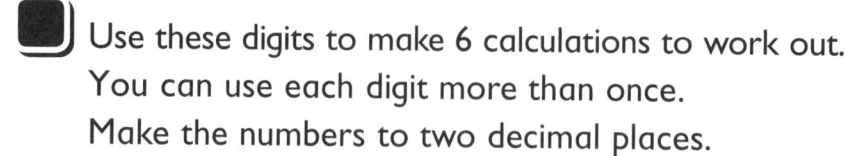

- Use efficient written methods to add decimals

■ Use these digits to make 6 calculations to work out.
You can use each digit more than once.
Make the numbers to two decimal places.
Then complete each calculation.

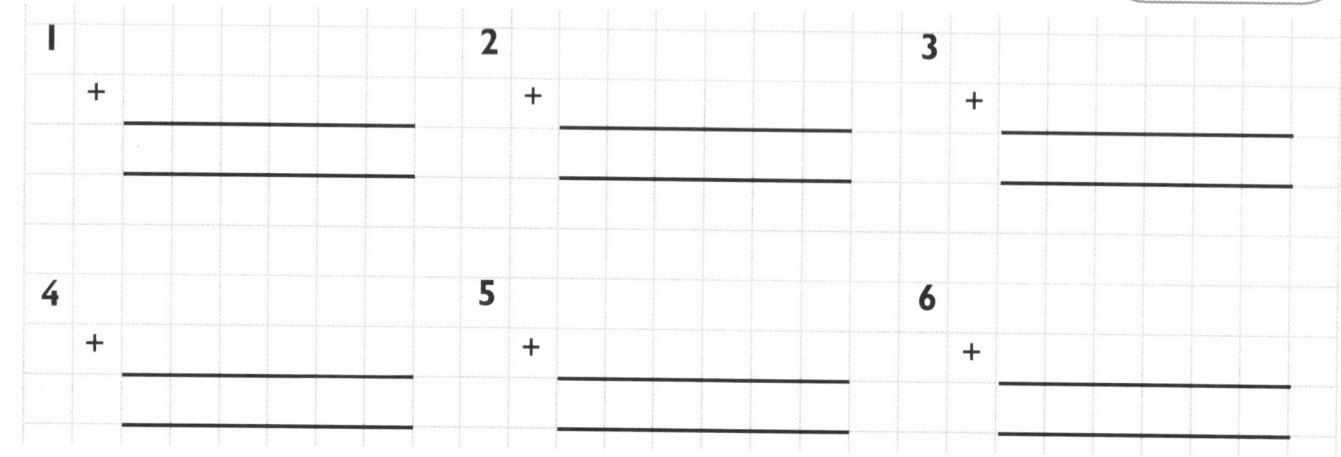

8 6 2
4 3 7

Example
```
  86.24
+ 38.47
———————
```

1   +
2   +
3   +
4   +
5   +
6   +

● Use these digits to make 6 calculations to work out.
Make the numbers in each calculation to a different number of decimal places.
Then complete each calculation.

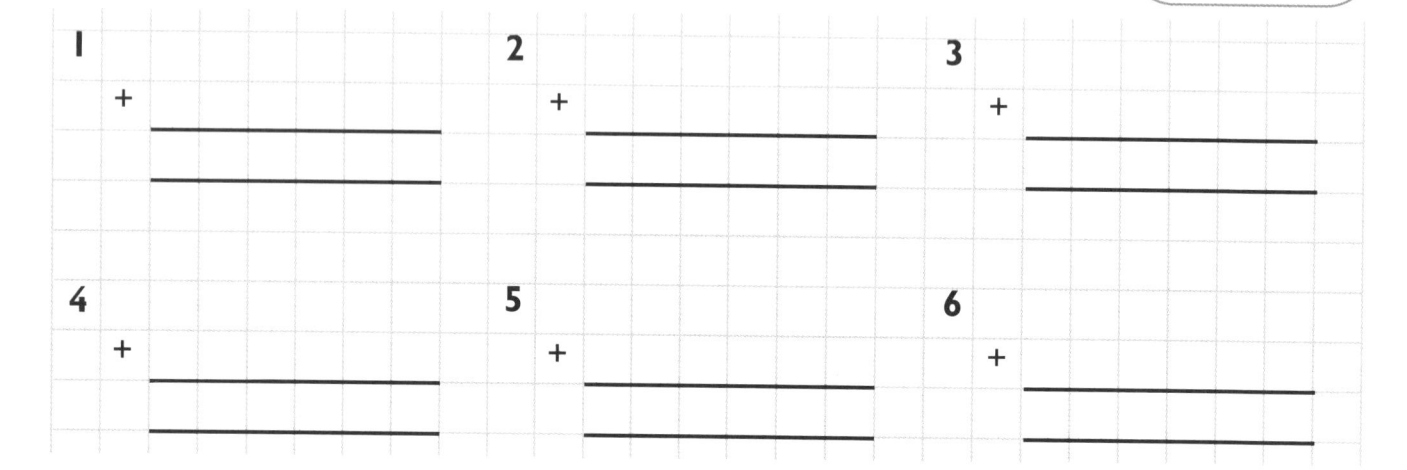

5 7 3
1 6 9

Example
```
   5.763
+ 67.319
————————
```

1   +
2   +
3   +
4   +
5   +
6   +

▲ Find the smallest and largest decimal with three places you can make using all the digit cards in the ● section in each number.
What is the total? ☐ What is the difference? ☐
Show your working out on the back of this sheet.

Y6 A3 HCM

Name _____ Date _____

Y6 A3 L6
HCM 51

Recording division

- Approximate first. Use efficient written methods to divide whole numbers

Approximate the answer to each calculation.

Example
465 ÷ 5 ≈ 450 ÷ 5 = 90

a	356 ÷ 4 ≈	b	263 ÷ 7 ≈
c	483 ÷ 5 ≈	d	662 ÷ 8 ≈
e	574 ÷ 9 ≈	f	285 ÷ 3 ≈
g	756 ÷ 8 ≈	h	526 ÷ 6 ≈

For each calculation above, work out the answer using a written method.

a 4)356

Answer =

b 7)263

Answer =

c 5)483

Answer =

d 8)662

Answer =

e 9)574

Answer =

f 3)285

Answer =

g 8)756

Answer =

h 6)526

Answer =

© HarperCollinsPublishers Ltd 2007

Long division

- Use efficient written methods to divide whole numbers

Work out the answers to these in your head.

a 360 ÷ 12 =
b 480 ÷ 24 =
c 750 ÷ 25 =
d 460 ÷ 23 =

e 390 ÷ 13 =
f 660 ÷ 22 =
g 720 ÷ 36 =
h 540 ÷ 27 =

i 450 ÷ 15 =
j 640 ÷ 16 =
k 840 ÷ 21 =
l 990 ÷ 33 =

Approximate the answer first. Then use a written method to work out the answers.

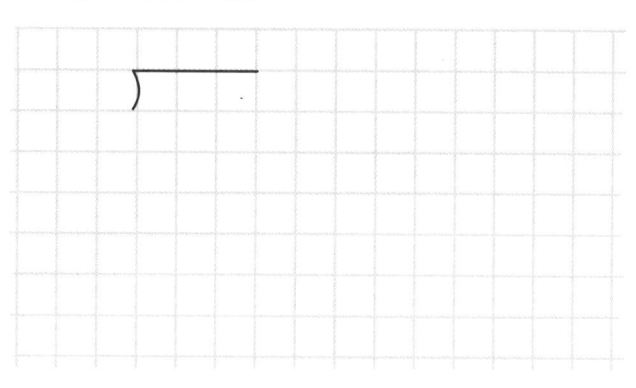

a 874 ÷ 23 ≈

23)874

Answer =

b 992 ÷ 23 ≈

Answer =

c 910 ÷ 26 ≈

Answer =

d 966 ÷ 42 ≈

Answer =

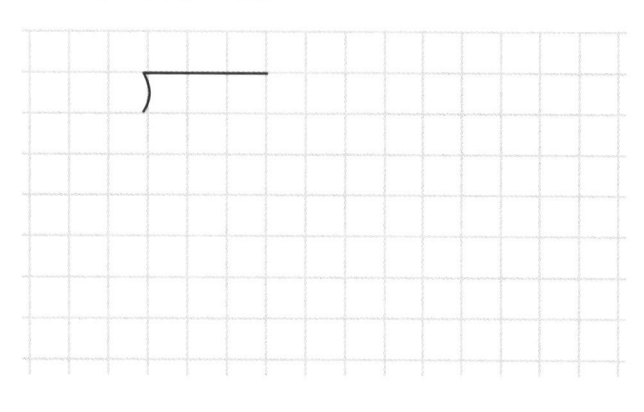

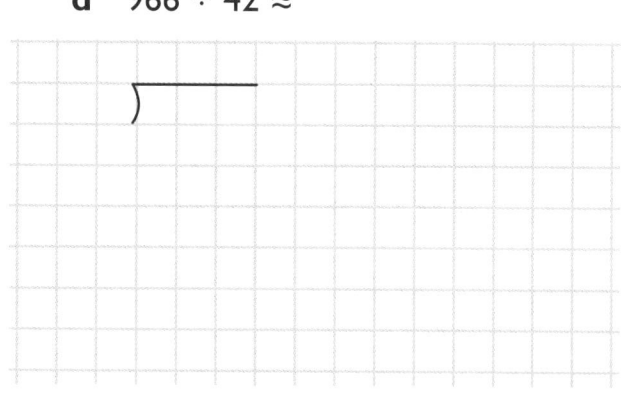

Y6 A3 HCM

Multiplying mentally by 11

- Use multiplication facts to work out mentally other facts

1 Carole is working out these examples for homework in this way:
 a 21 × 11 = (21 × 10) + (21 × 1) = 210 + 21 = 231

Complete examples **b** to **d** in the same way.
 b 22 × 11 = (   ) + (   ) =   +   =
 c 23 × 11 = (   ) + (   ) =   +   =
 d 24 × 11 = (   ) + (   ) =   +   =

2 She spots a pattern.

I need to work out 25 × 11.

I'll write 2 ... 5 and fill the space with 2 + 5 = 7.

The answer is 275.

I'll check to be sure.
 25 × 10 = 250
 25 × 1 = + 25
                275

Use Carole's quick way to work out these mentally.
 a 26 × 11 =
 b 27 × 11 =
 c 32 × 11 =
 d 53 × 11 =

I'm stuck at 29 × 11. 2119 is too big.

I'll check.
 29 × 10 = 290
 29 × 1 = 29
                319

Aha! 2 tens and 9 tens is 11 tens which is 1 hundred and 1 ten.

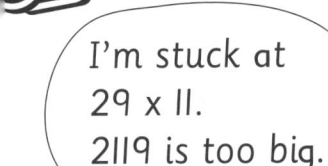

The quick way works!
   2 9
 + 1 1 0
   3 1 9

 a 38 × 11 =   d 69 × 11 =
 b 46 × 11 =   e 77 × 11 =
 c 58 × 11 =   f 98 × 11 =

Y6 B3 L1
HCM 53

© HarperCollinsPublishers Ltd 2007

Name _____ Date _____

Tests of divisibility (2)

- Use approximations, inverse operations and tests of divisibility to estimate and check results

1 In the number grid, colour half of each square where the number is divisible by 3 or 4.

Key blue – divisible by 3
  red – divisible by 4

You need:
- red and blue pencil

2 List the numbers that are divisible by 3 and 4.

3 List the numbers that are divisible:

  a by 6
  b by 8
  c by 6 and 8

401	402	403	404	405	406
407	408	409	410	411	412
413	414	415	416	417	418
419	420	421	422	423	424
425	426	427	428	429	430
431	432	433	434	435	436
437	438	439	440	441	442
443	444	445	446	447	448
449	450	451	452	453	454
455	456	457	458	459	460

1 Complete the table.

	÷ 3	÷ 4	÷ 5	÷ 6	÷ 7	÷ 8	÷ 9	÷ 10	÷ 12	÷ 25
408										
432										
441										
450										

2 Find 2 different numbers in the first table that are divisible by 5 and 7.

3 Find 2 different numbers in the table that are divisible:

  a by 11 ___ , ___ , **b** by 15 ___ , ___

Y6 B3 HCM

Multiples and consecutive numbers

- Describe and explain sequences and relationships

You can write the multiples of 3 as the addition of three consecutive numbers.

15 = 5 + 5 + 5
   = (5 − 1) + 5 + (5 + 1)
   = 4 + 5 + 6

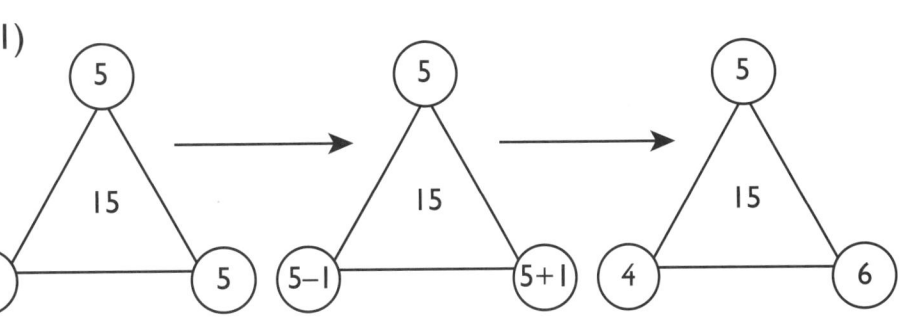

Find three consecutive numbers which total the numbers in the triangles.

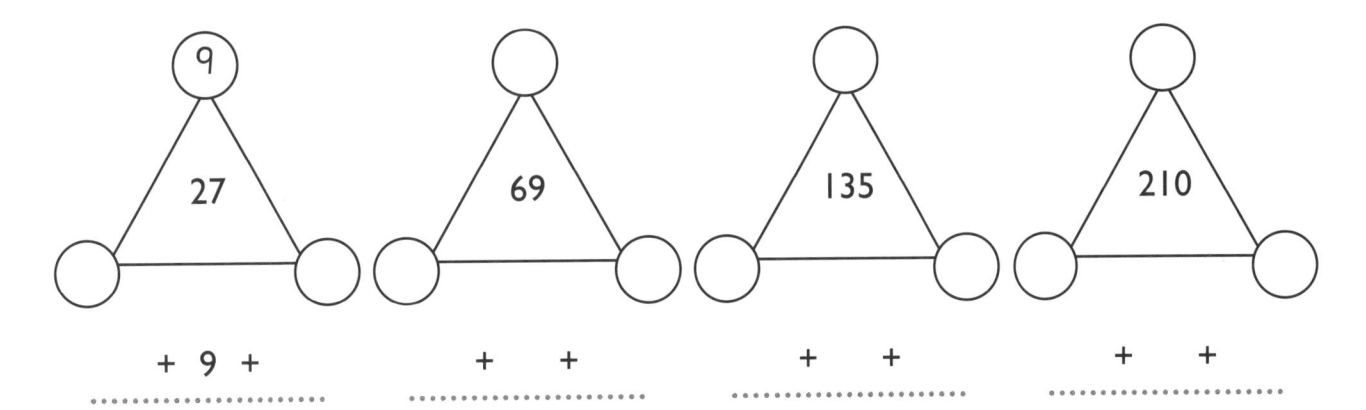

Triangles: 27 (top 9), 69, 135, 210

+ 9 + ...

+ + ...

+ + ...

+ + ...

Find five consecutive numbers which total the multiples of 5.

Complete the table.

Multiple of 5	5 consecutive numbers
15	1 + 2 + 3 + 4 +
20	+ + 4 + +
25	+ + + +
65	+ + + +
90	+ + + +
125	+ + + +
200	+ + + +

Y6 B3 L7
HCM 55

© HarperCollinsPublishers Ltd 2007

Agent P's puzzle

- Find which numbers less than 100 are prime

The flagstones on the floor of an ancient castle are marked with a spiral of numbers. Agent P must find the safe route to the flagstone marked 1 to recover the microfilm.

You need:
- colouring materials

Instructions
- The safe stones are marked with prime numbers.
- Begin at a prime number greater than 64.
- Move diagonally only until you reach 1.
- Your stones are in descending order.
- Avoid all non-prime stones.

NORTH

73	74	75	76	77	78	79	80	81	82
72	43	44	45	46	47	48	49	50	83
71	42	21	22	23	24	25	26	51	84
70	41	20	7	8	9	10	27	52	85
69	40	19	6	1	2	11	28	53	86
68	39	18	5	4	3	12	29	54	87
67	38	17	16	15	14	13	30	55	88
66	37	36	35	34	33	32	31	56	89
65	64	63	62	61	60	59	58	57	90
100	99	98	97	96	95	94	93	92	91

WEST / **EAST**

SOUTH

1. Shade in the prime numbers in the grid.

2. List the 3 possible safe routes for Agent P.

   Route 1 ..... → ..... → ..... → ..... → ..1..

   Route 2 ..... → ..... → ..... → ..... → ..1..

   Route 3 ..... → ..... → ..... → ..... → ..1..

Any number having only two factors, itself and 1, is a prime number.

1 is not a prime number.

3. Agent P's final instruction reads:

   The total of the prime numbers for the safe route is 156. From which direction must he begin?

Y6 B3 HCM

Y6 B3 L12
HCM 57

Name _____ Date _____

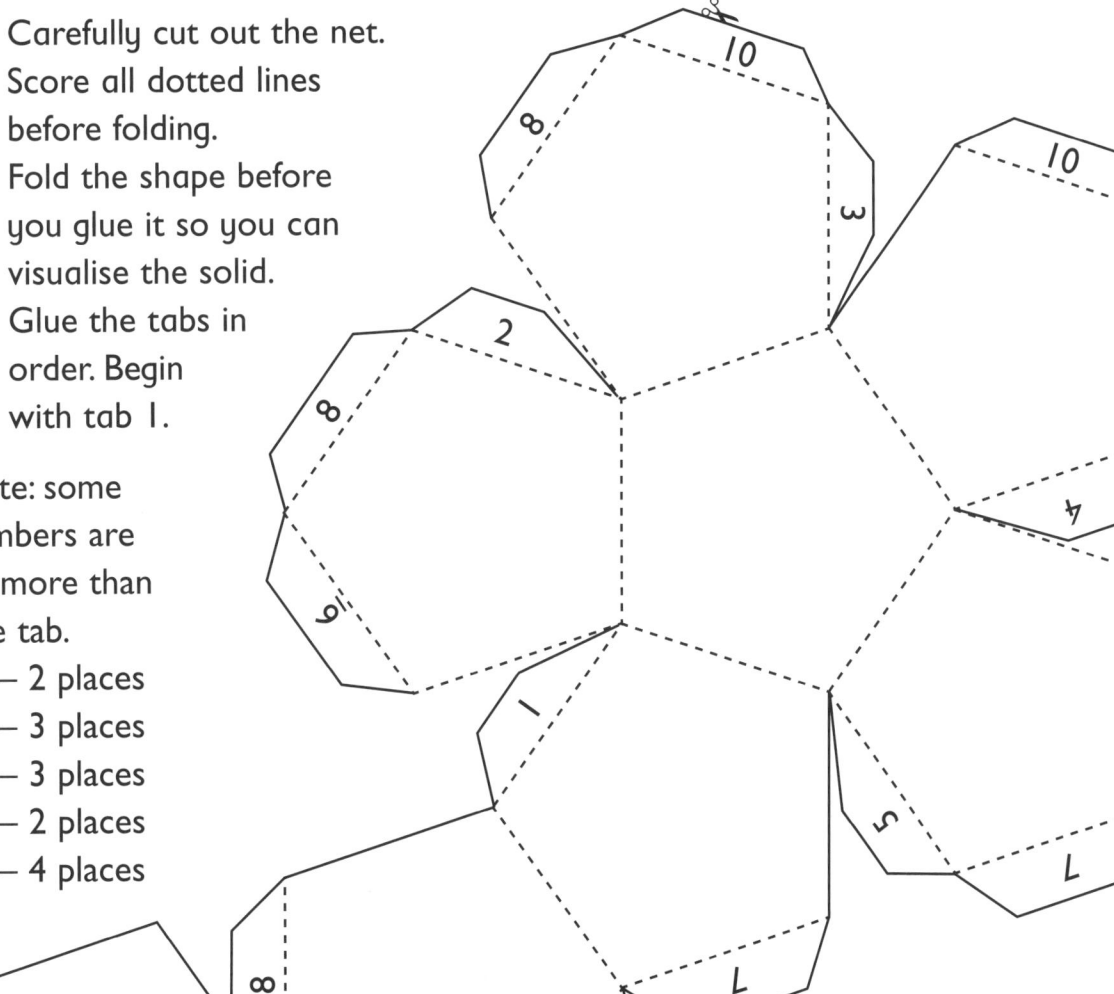

Making a dodecahedron

- Make shapes with increasing accuracy

1. Carefully cut out the net.
2. Score all dotted lines before folding.
3. Fold the shape before you glue it so you can visualise the solid.
4. Glue the tabs in order. Begin with tab 1.

Note: some numbers are on more than one tab.
- 6 – 2 places
- 7 – 3 places
- 8 – 3 places
- 9 – 2 places
- 10 – 4 places

You need:
- scissors
- ruler
- glue

© HarperCollinsPublishers Ltd 2007

Collins New Primary Maths

Y6 B3 L
HCM

Name _____ Date _____

Grid patterns

- **Make and draw shapes with increasing accuracy**

1. Make 2 different rotating patterns, one on each grid.
2. Use no more than three colours to highlight your patterns.
3. Continue each pattern as far as you can go.

You need:
- colouring materials

© HarperCollinsPublishers Ltd 2007

Name _____ Date _____

Converting capacities

- **Convert smaller units to larger units and vice versa**

Y6 C3 L2
HCM 59

You have four containers (A to D) and a large bottle (E).

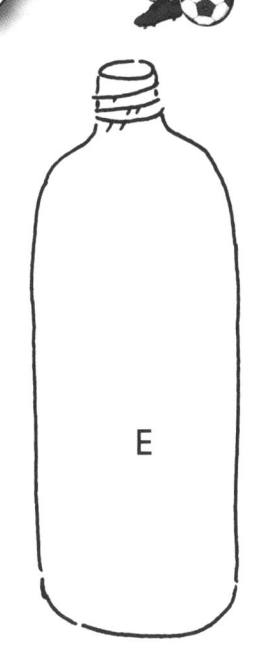

A 100 ml   B 75 cl   C 0·25 l   D 1½ l   E

1 Write the capacity of each container:

container	millilitres	centilitres	litres
A	100 ml		
B	750 ml	75 cl	
C			0·25 l
D			

2 Show how you can use the containers to fill bottle E with these amounts of water.

Amount in E	Containers used	Check
a 1 litre	B and C	750 ml + 250 ml = 1000 ml = 1 l
b 0·350 l		
c 0·6 l		
d 85 cl		
e 1·25 l		

Find the capacity in litres of bottle E if it holds:

a 2 of B + 2 of C + 1 of D ………… litres

b 1 of A + 2 of C + 3 of D ………… litres

c 3 of B + 4 of D ………… litres

d 5 of A + 4 of B + 3 of C ………… litres

© HarperCollinsPublishers Ltd 2007

Currency converter

- Use a graph to convert money

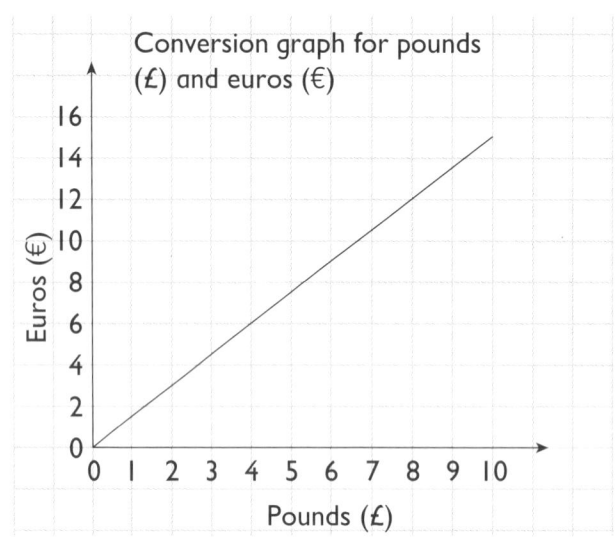

You need:
- graph paper

Conversion graph for pounds (£) and euros (€)

1 Estimate these amounts in euros (€). Use the graph to help you.

a £4 ___   b £10 ___   c £6 ___   d £3.50 ___

2 Use the conversion graph to convert these amounts to pounds (£).

a €12 ___   b €3 ___   c €10 ___

Converting to Australian Dollars (AU$)

Pounds (£)	Australian Dollars (AU$)
0	0
40	100
80	200

1 Copy and complete the conversion graph.

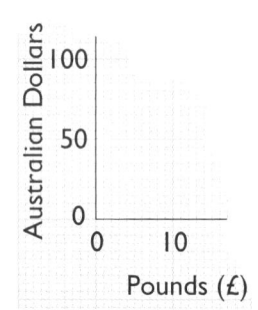

2 Use your graph to complete this bill.

STAR HOTEL

Item	£	AU$
Room	70	
Meals		130
Drinks	14	
Entertainment		45
Phone calls	6	
Laundry		20

£80 is worth around 1200 Hong Kong Dollars (HK$).

1 Draw a graph to convert pounds (£) to Hong Kong Dollars (HK$).

2 Convert **a** £50 to Hong Kong Dollars and **b** HK$900 to pounds.

Y6 C3 L7
HCM 61

Name _____ Date _____

Musical pie charts

- Represent data in different ways and understand its meaning

Four music bands touring the country together sold their own CDs. The pie charts show their sales in Birmingham and Glasgow.

Birmingham

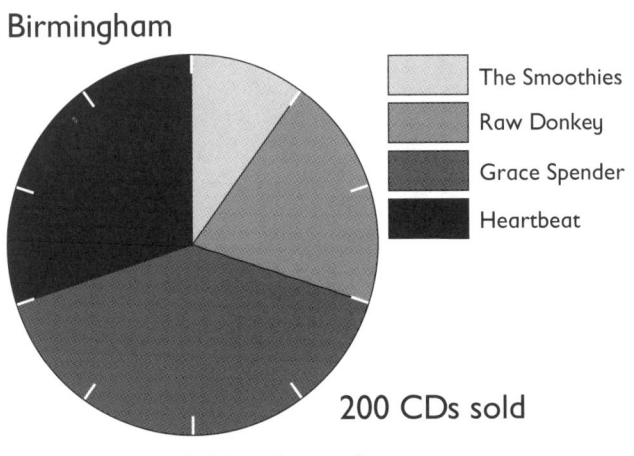

- The Smoothies
- Raw Donkey
- Grace Spender
- Heartbeat

200 CDs sold

Glasgow

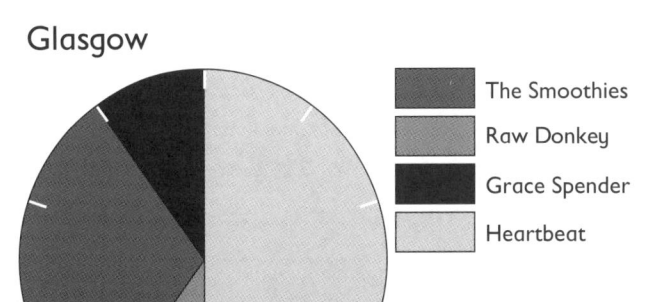

- The Smoothies
- Raw Donkey
- Grace Spender
- Heartbeat

Band	Number of CDs sold
Total	

Band	Number of CDs sold
Total	

1 Complete the table for Birmingham.

2 Raw Donkey sold 16 CDs in Glasgow.
   a What percentage did Raw Donkey sell? ☐
   b How many CDs were sold altogether? ☐
   c Complete the table for Glasgow.

3 Write two sentences comparing the sales in the two cities.

You need:
- Small circular object, e.g. rim of a cup
- protractor
- ruler
- colouring materials

Rachel's mp3 player has 30 pop, 20 classical, 40 rap and 10 jazz tracks.

Draw a pie chart for this data on the back of this sheet.

© HarperCollinsPublishers Ltd 2007

Y6 C3
HCM

Name _____ Date _____

Holiday bar charts

- Represent data in different ways and understand its meaning

■ The numbers show the days of holiday some people took last year.

14 25 7 33 43 21	35 49 18 28 14 42	35 36 28 14 18 35
11 6 14 18 38 28	35 26 28 21 40 28	21 12 28 30 38
35 30 21 48 42	7 30 42 42	19 28 42 49 9 42

1 Record the data in the tally and frequency columns on this table.
Which do you think will be the most popular class? ....

Days	Tally	Frequency	Percentage
0–9			
10–19			
20–29			
30–39			
40–49			
	Total		

2 a How many people took fewer than 20 days of holiday? ....
  b How many people took more than 29 days of holiday? ....
  c How many people took from 10 to 29 days of holiday? ....

● 1 Calculate the total frequency and write it in the table.
  2 Convert the frequencies to percentages. Use your calculator.
  3 Complete this percentage bar chart.
  4 Was your prediction in Q1 correct? ☐

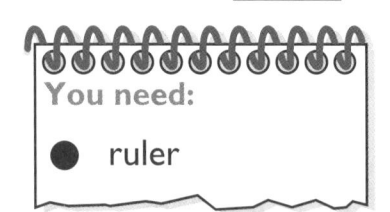

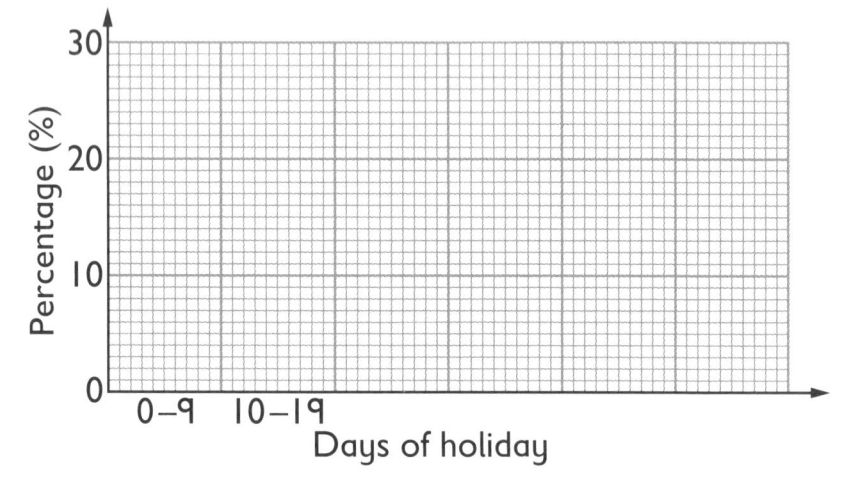

You need:
- ruler

▲ On the back of this sheet calculate the mean, median, mode and range of the data in ■

© HarperCollinsPublishers Ltd 2007

Litres and gallons

- Know imperial units (gallons)
- Know rough equivalents of litres and gallons

Remember
4·5 litres ≈ 1 gallon

1. Complete the conversion scale for litres and gallons.
2. Find the tank capacity in litres, to the nearest litre, for each aquarium in the pet shop.

a 5·5 gallons ☐ l

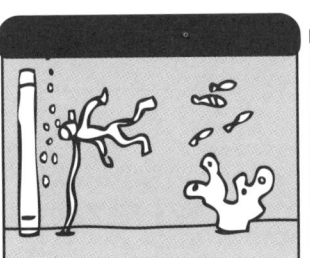

b 3·3 gallons ☐ l

c 6·6 gallons ☐ l

d 7·9 gallons ☐ l

e 8·6 gallons ☐ l

f 4·6 gallons ☐ l

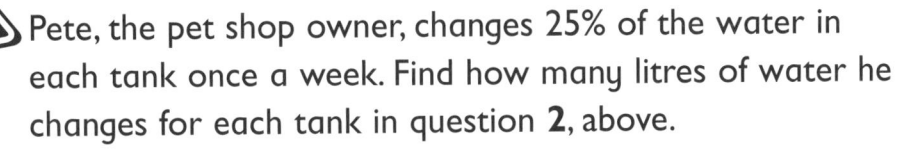

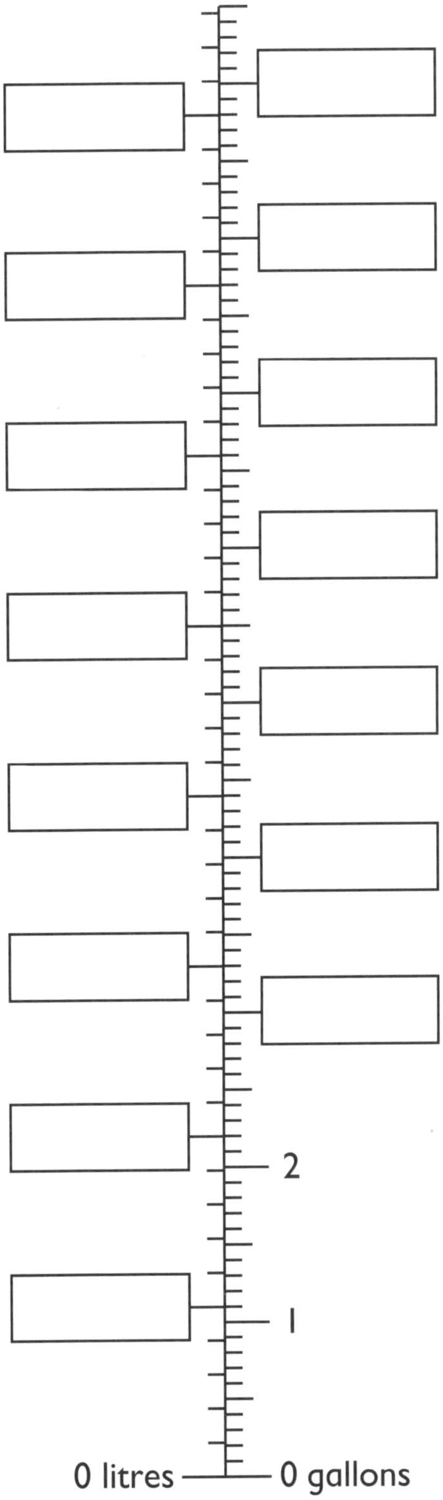

Pete, the pet shop owner, changes 25% of the water in each tank once a week. Find how many litres of water he changes for each tank in question **2**, above.

Tank	Capacity (litres)	25% of capacity
a		
b		
c		
d		
e		
f		

0 litres — 0 gallons

Pinboard areas

- Calculate the area of a shape formed from rectangles

You can work out the area of a shape made on a pinboard in 2 ways:

1 The sum of smaller shapes

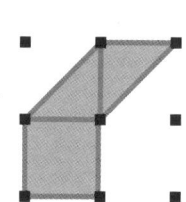

2 What you have left when you cut pieces off a larger shape

Find the area in square centimetres of these shapes.

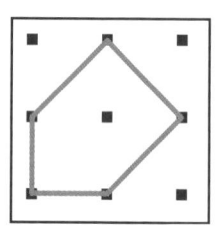

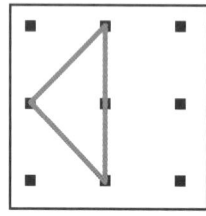

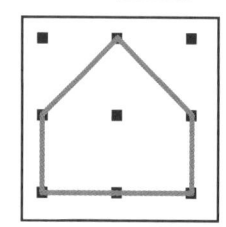

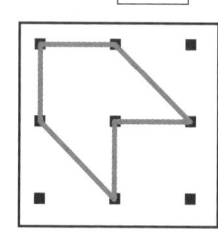

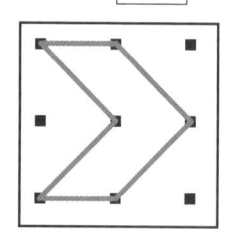

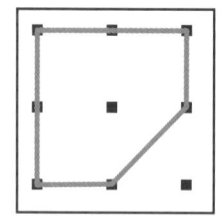

area = ☐ cm²   area = ☐ cm²   area = ☐ cm²   area = ☐ cm²

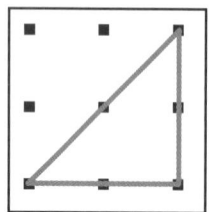

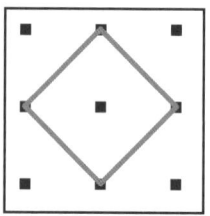

area = ☐ cm²   area = ☐ cm²   area = ☐ cm²   area = ☐ cm²

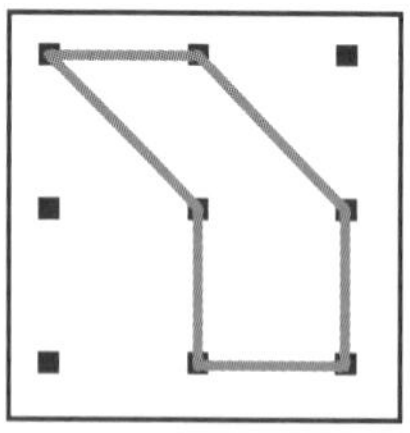

area = ☐ cm²   area = ☐ cm²   area = ☐ cm²   area = ☐ cm²

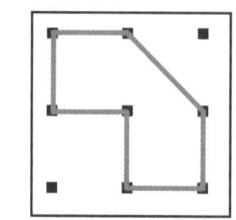

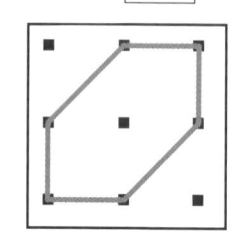

Work out the area of the shapes in **a** and **b** below. Draw shapes with the areas given in **c** and **d**.

You need:
- ruler

a   area = ☐ cm²
b   area = ☐ cm²
c   area = 2·5 cm²
d   area = 3·5 cm²

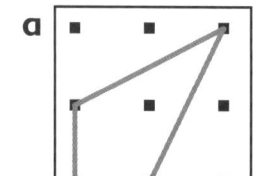

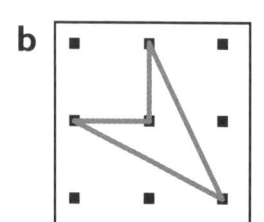

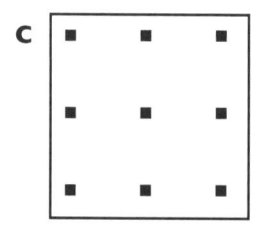

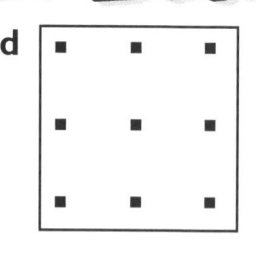

Finding surface areas

- Find the surface area of a cuboid

A farmer fences off a rectangular field for his pigs and poultry. He leaves the rest of the field for his sheep. Find the area of ground in square metres for:

a pigs   Area = ... m²

b poultry   Area = ... m²

c sheep   Area = ... m²

working out

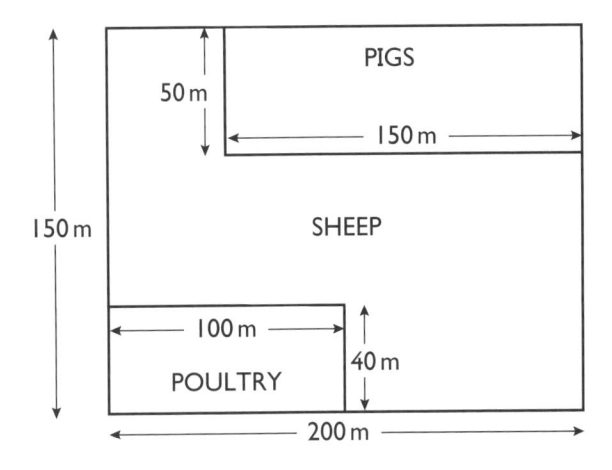

This set of cuboids is made of centicubes.

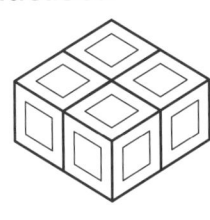

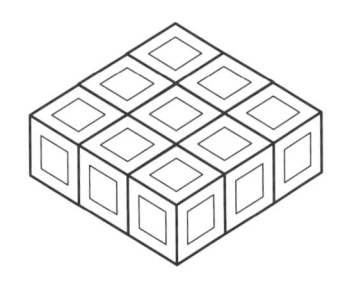

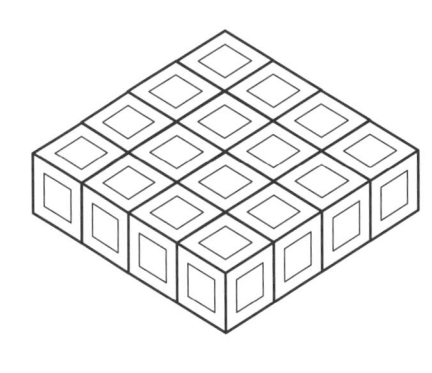

A   B   C   D

1 Work out the surface area of each cuboid. Record your answers in the table.

Cuboid	A	B	C	D	E	F
Length of side (s)	cm	cm	cm	cm	cm	cm
Surface area (A)	cm²	cm²	cm²	cm²	cm²	cm²

2 Find a pattern and use it to find the surface area of the next two cuboids, E and F, in the sequence.

3 A cuboid in the sequence has a length of side of 10 centicubes.

Calculate its surface area.   Surface area = ... cm².

Checking calculations

- Use approximations and inverse operations to estimate and check results

Make up 12 addition or subtraction calculations:

4 you can do in your head;

4 to do on a calculator;

4 to do using the written method.

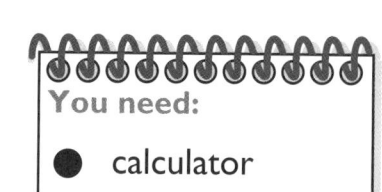

You need:
- calculator

Choose a way to check each calculation. This may be using the inverse operation, estimating first or using a different method to work it out.

Calculation	Working out	Checking

Explain why it is important to check answers to calculations.

....

....

....

Decimal fraction pairs

- Find equivalent decimals and fractions

Make a set of fraction and decimal cards. Use the cards to play Pairs.
- Write these fractions on half of the cards and their decimal equivalent on the other half of the cards.

■ $\frac{1}{2}, \frac{1}{4}, \frac{3}{4}, \frac{1}{5}, \frac{3}{5}, \frac{1}{8}, \frac{5}{8}, \frac{1}{10}, \frac{4}{10}, \frac{1}{3}$

● $\frac{7}{10}, \frac{3}{8}, \frac{3}{4}, \frac{1}{4}, \frac{1}{3}, \frac{1}{6}, \frac{2}{3}, \frac{4}{5}, \frac{7}{8}, \frac{6}{8}$

HINT
Only use the calculator if you need to.

- Lay the cards face down on a table.
- Turn over two cards. If you are playing with someone else, take it in turns to choose two cards.
- If the fraction and decimal are equivalent, keep them. If not, put them back in the same place.
- Keep going until you have made all the pairs.

You need:
- scissors
- calculator (optional)

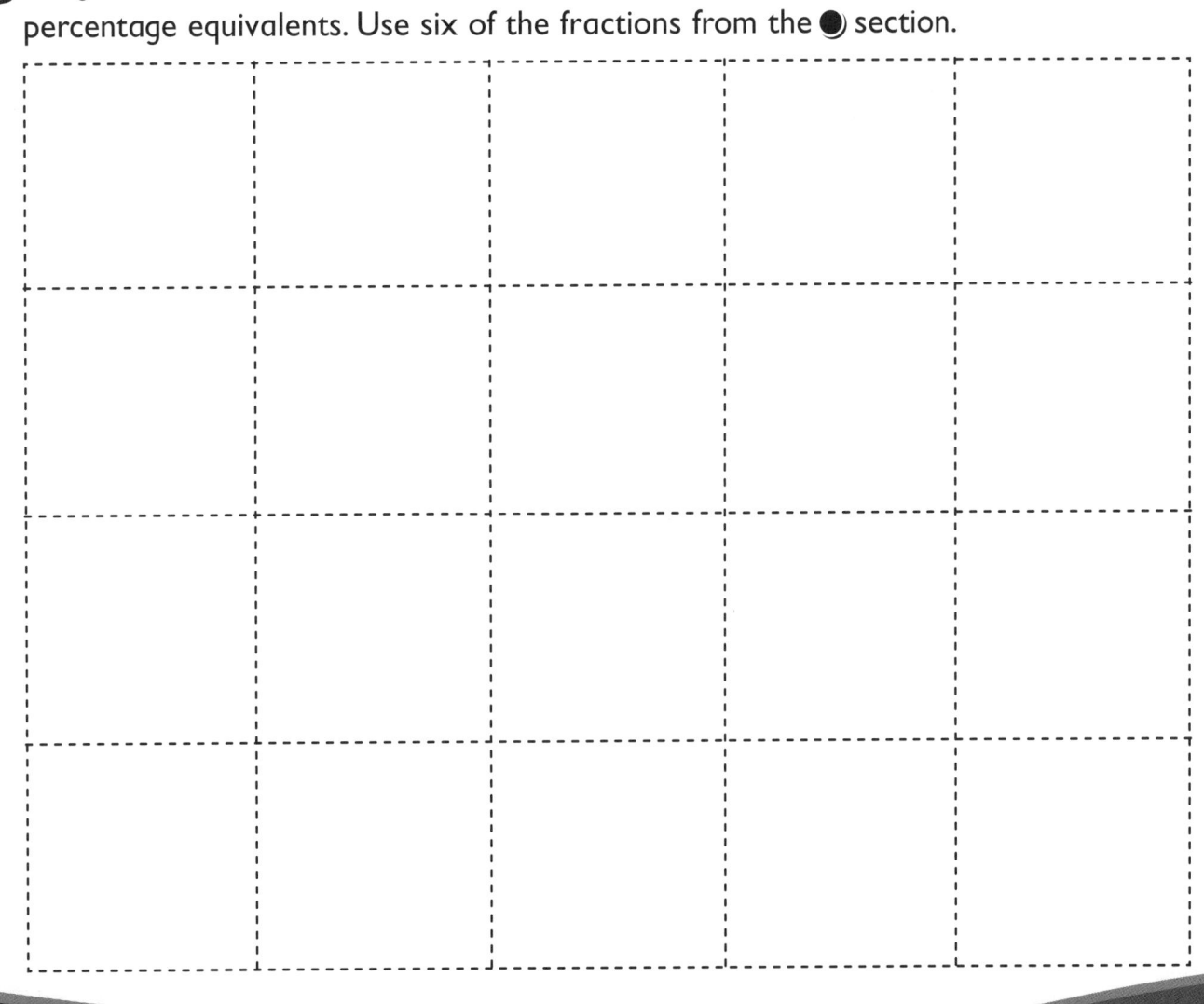

▲ Only use 18 cards, and make a set of cards that have fraction, decimal and percentage equivalents. Use six of the fractions from the ● section.

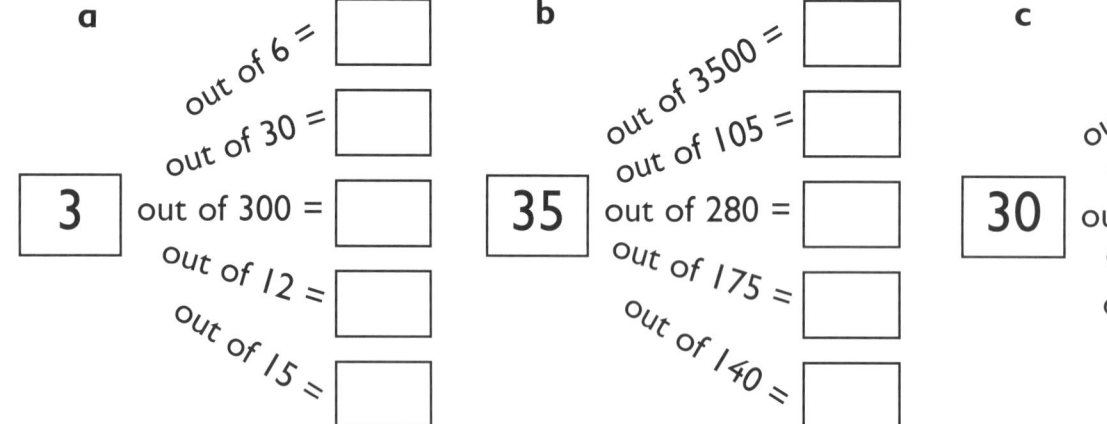

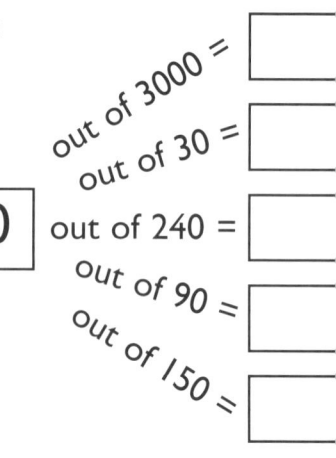

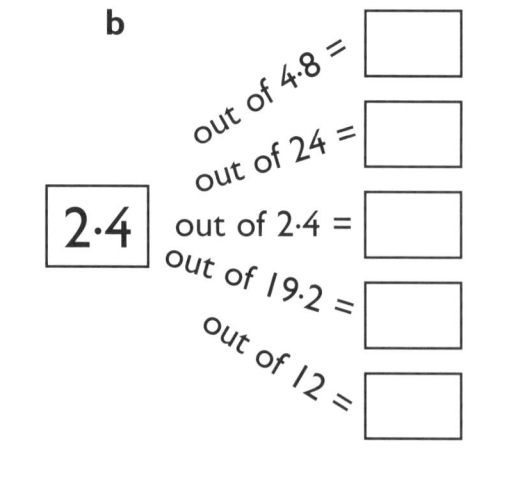

Percentages

- **Express one amount as a percentage of another**

Work out the first number as a percentage of the second number.

Example: 2 out of 8 — 2 is 25% of 8

a. 20 out of 40 — 20 is [ ] % of 40
b. 5 out of 25 — 5 is [ ] % of 25
c. 4 out of 16 — 4 is [ ] % of 16
d. 6 out of 60 — 6 is [ ] % of 60
e. 3 out of 60 — 3 is [ ] % of 60
f. 9 out of 90 — 9 is [ ] % of 90
g. 27 out of 90 — 27 is [ ] % of 90
h. 12 out of 120 — 12 is [ ] % of 12
i. 48 out of 120 — 48 is [ ] % of 120
j. 9 out of 45 — 9 is [ ] % of 45

Work out the first number as a percentage of the other numbers.

a. 3
- out of 6 = [ ]
- out of 30 = [ ]
- out of 300 = [ ]
- out of 12 = [ ]
- out of 15 = [ ]

b. 35
- out of 3500 = [ ]
- out of 105 = [ ]
- out of 280 = [ ]
- out of 175 = [ ]
- out of 140 = [ ]

c. 30
- out of 3000 = [ ]
- out of 30 = [ ]
- out of 240 = [ ]
- out of 90 = [ ]
- out of 150 = [ ]

a. 4·5
- out of 9 = [ ]
- out of 18 = [ ]
- out of 22·5 = [ ]
- out of 45 = [ ]
- out of 450 = [ ]

b. 2·4
- out of 4·8 = [ ]
- out of 24 = [ ]
- out of 2·4 = [ ]
- out of 19·2 = [ ]
- out of 12 = [ ]

Y6 E3 HCM

Y6 E3 L8
HCM 69

Name _____ Date _____

Addition and subtraction

- Use efficient written methods to add and subtract whole numbers and decimals

Example
$$6 + 4 = 10$$
$$10 - 6 = 4$$

■ Fill in the third number in each triangle. The three numbers go together to make an addition and subtraction fact. No number will be more than 20.

Write the two facts underneath.

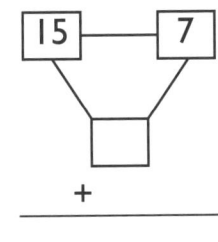

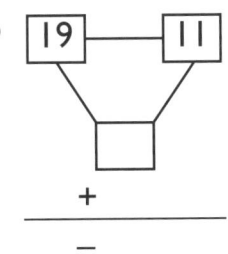

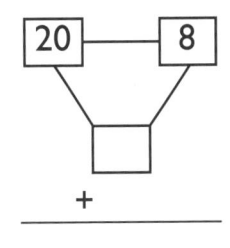

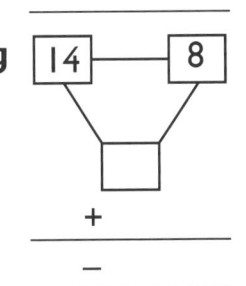

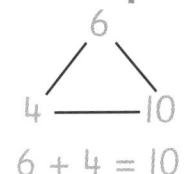

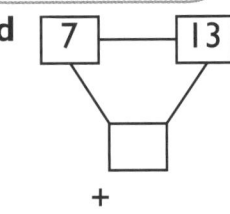

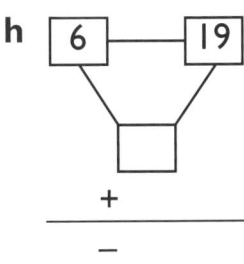

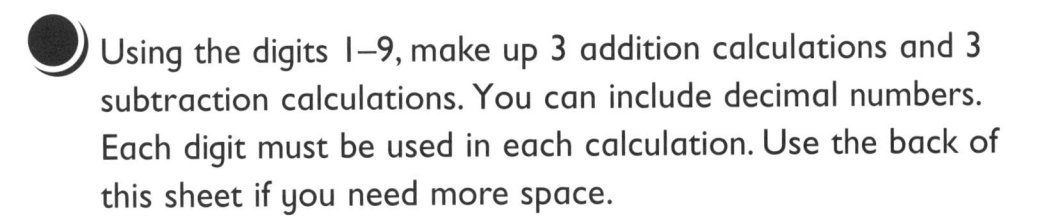

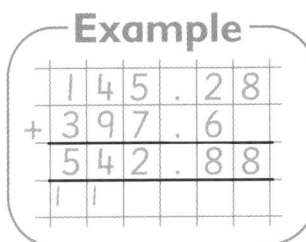

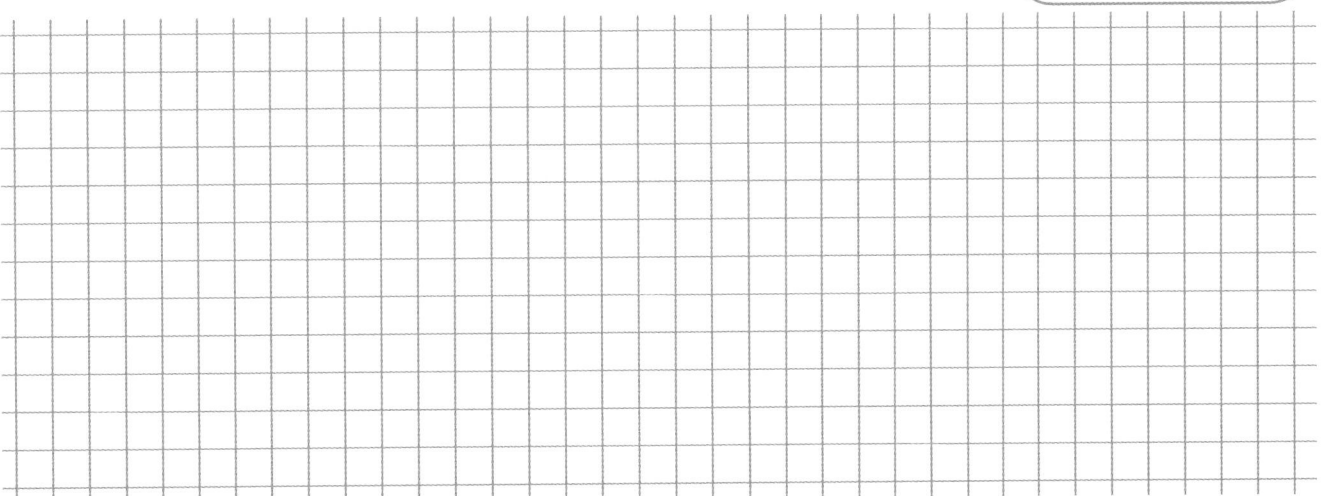

a) 15 — 7

b) 19 — 11

c) 20 — 8

d) 7 — 13

e) 12 — ☐, 17

f) 16 — 8

g) 14 — 8

h) 6 — 19

● Using the digits 1–9, make up 3 addition calculations and 3 subtraction calculations. You can include decimal numbers. Each digit must be used in each calculation. Use the back of this sheet if you need more space.

Example
$$145.28 + 397.6 = 542.88$$

▲ What is the smallest answer you can get? What is the largest answer you can get? Show your working on the back of this sheet.

© HarperCollinsPublishers Ltd 2007

Name _____ Date _____

Ratio investigation

- Solve simple problems involving ratio

The ratio of the distance across a circle to the distance around the edge of it is approximately 1 to 3.
Find out if this statement is true by measuring 10 circular items in your home.

You need:
- tape measure or string / wool
- ruler

Object	Distance across circle	Distance around circle	Ratio

Y6 E3 HCM

Y6 E3 L12
HCM 71

Name _____ Date _____

Solving word problems

- Choose and use appropriate number operations to solve word problems involving money

■ Solve each word problem. Show all your working.

a Buy 3 squash rackets and 1 pair of ice skates. What is the total cost?

b Buy 2 snowboards. How much change do you receive from £200?

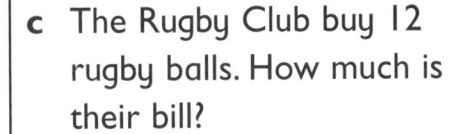

c The Rugby Club buy 12 rugby balls. How much is their bill?

d Top of the range snow skis cost 3 times more than the pair shown here. How much do they cost?

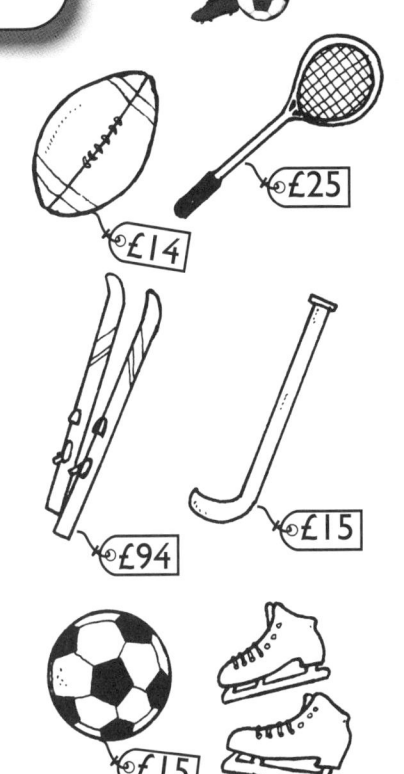

Prices shown: £25, £14, £94, £15, £15, £24, £78

■ Solve each word problem. Show all your working.
Do your working on the back of this sheet if you need more room.

a The ski-hire shop needs new equipment. The owner buys 8 snowboards, 9 pairs of skis and 15 pairs of ice skates. What is the total cost of the purchases?

b The sports store has 45 squash rackets to sell. The owner paid about £1000 to the factory for the rackets. How much profit will they make if all rackets are sold?

c During the sale, there is a free football for every snowboard sold. The shop sells 24 snowboards. How much money is made altogether?

d The local sports club buys 7 rugby balls, 12 footballs and 23 squash rackets. How much change is there from £1000?

© HarperCollinsPublishers Ltd 2007

Reflective bunting

- Record the information in a problem or puzzle

■ Year 6 is making bunting in the school colours of blue and gold for the summer fete.

Each piece of bunting is made up of 4 small triangles.

How many different types of bunting can they make using only two colours?

You need:
- blue and yellow colouring materials
- ruler

● Complete this table.

Colours	Number of pieces
2 blue and 2 gold	
1 blue and 3 gold	
3 blue and 1 gold	

▲ All the flags have reflective symmetry.
Mark each line of symmetry with a dotted line.

Y6 E3 L HCM